Divine Games

Divine Games

Game Theory and the Undecidability of a Superior Being

Steven J. Brams

The MIT Press
Cambridge, Massachusetts
London, England

This book was set in Palatino LT Std by Toppan Best-set Premedia Limited.

Library of Congress Cataloging-in-Publication Data is available.

ISBN: 978-0-262-03833-1 (hc)
ISBN: 978-0-262-55145-8 (pb)

I confess that at times I have been tempted to believe that the Creator has eternally intended ... to remain *baffling*, to prompt our curiosities and hopes and suspicions all in equal measure. ... Although ... spirits are always seeming to exist and can never be fully explained away, they also can never be fully susceptible to full corroboration.

—William James (1909)

Contents

Preface

In this book I analyze several decisions and games that a human being, called P (for "person"), might play with a godlike being, called SB (for "superior being"). I mix the theoretical analysis of classical theological questions, raised by Pascal's wager and Newcomb's problem, with the analysis of several stories from the Hebrew Bible, almost all of which involve conflict between God, or a surrogate, and a human player. Play of these games raises fundamental questions about P's relationship to SB, which is the main subject of theology.[1]

Although God's existence is almost never questioned in the Bible, His powers certainly are. For many people today, however, whether God or any SB exists remains an open question. So does the role that God (SB) plays if He (he) does exist.

To be sure, SB's existence is *not* an open question for either believers (theists) or nonbelievers (atheists). While theists believe there is ample, if not compelling, evidence for the existence of SB, atheists find no such evidence. Indeed, many people believe that natural forces, without SB, perfectly well explain the formation of the universe over billions of years and the evolution of human beings on earth.

If you are neither a theist nor an atheist, by default you are an agnostic, believing the question of SB's existence—or any supernatural phenomena (e.g., perhaps multiple gods)—is not yet settled, either affirmatively or negatively. In this book, I give reasons why this question is likely to remain unsettled, but not just because it is a default position. Instead, I argue that it is difficult, if not impossible, to

---

1. I make P and SB the players in the biblical games to facilitate the comparison of these games in which the cast of human players changes but SB is always either God or a human player acting in God's stead. I use "He" to refer to God in the Bible stories, because this is standard practice in traditional texts, and "he" (without capitals) for SB. P is "she," but I hasten to add that these male and female pronouns for SB and P, respectively, could as well be reversed.

determine whether or not SB exists from the rational choices biblical characters made—and that we make today in games that we might play with SB.

My analysis of the outcomes of biblical games does not unequivocally establish SB's existence or—if he does exist—whether or not he is in any more control than P is. Therefore, it seems advisable to keep an open mind about SB's existence and the role he plays in human affairs. This is an argument for agnosticism, not by default but because SB's superiority is *undecidable*, or inherently unknowable.[2]

What light does game theory shed on this issue? To begin with, it predicts outcomes in games that are "in equilibrium," meaning roughly that they are stable and players would not want to depart from them. But there are different notions of equilibrium in game theory; I focus on those that are "nonmyopic," whereby farsighted players, thinking ahead, consider them stable. Even if they have an immediate incentive to move from them, they may not do so, according to the *theory of moves*, or TOM (Brams 1994), which is a cornerstone of the models I develop.[3]

TOM assumes that P and SB are capable of thinking ahead when they plot their moves and countermoves in a game. It provides a subtler notion of equilibrium than the standard notion in game theory, proposed by John Nash and called a "Nash equilibrium," whereby no player *immediately* benefits by moving from such an equilibrium, even though later moves may be beneficial.

TOM drastically changes the rules of play of standard game theory to allow for moves and countermoves from some initial state. These choices can be accounted for in game theory, but only by complicating the notion of a strategy in a payoff matrix.

I use TOM to determine not only the rational choices of P and SB but also to elaborate the dynamics of moves and countermoves they make to achieve a nonmyopic equilibrium in a game. I also show how the exercise of different kinds of power (moving, order, and threat) by SB, and sometimes P, can affect game outcomes.

---

2. In mathematical logic, "undecidable" means not soluble from the axioms of set theory, as first developed in the work of Kurt Gödel in the 1930s; see, for example, Goldstein 2006. However, the usage of the term here, as will be discussed later, is game-theoretic.

3. Although game theory and TOM are developed from scratch in this book, their applications require careful reading and study for those new to game theory. To facilitate the reader's understanding, I illustrate the concepts introduced and the calculations underlying them with numerous examples; I also include a Glossary of more technical terms for quick reference. Applications of TOM to the humanities generally, including literature and history, are given in Brams 2011.

In general, the resulting outcomes do not establish that SB, even with his supernatural powers of omniscience and omnipotence, can always achieve his goals. Complicating matters, there may be multiple non-myopic equilibria, depending on where play of a game starts.

Paradoxically, these equilibria may not always benefit SB, despite his ability to foretell the future and exercise some kind of power over P. Thereby the effects of SB's superiority are rendered murky, making ambiguous, if SB exists, whether he is in fact superior or even in control.

Evidence that the biblical God is not always in control can be found in numerous games that He plays, including those with His chosen people, the Israelites, who on occasion flout His commands. They are usually punished, but not always, or not to the degree that God threatens. And sometimes those who sin escape punishment entirely for their crimes, including murder. At other times, the righteous are severely punished, occasionally for no apparent reason.

These events in the Bible make God's choices seem arbitrary, sometimes even random. They offer a possible explanation for the existence of evil in the world, because the wicked are not always punished nor the righteous rewarded. What further muddies the waters is that human players in the Bible almost always exercise free will, which makes their choices hard to predict, even for God.

In the world today, whether SB is a player, much less one with supernatural powers like omnipotence or omniscience, is even harder to determine. Consequently, SB's existence appears *undecidable*: Even if SB is able to influence human choices, any rhyme or reason behind them may be difficult to discern. Moreover, these choices do not seem rooted in any consistent code of ethics that distinguishes good from evil.

True, this does not prove that SB does not exist; neither does it prove that he exists. Rather, it shows that his actions are not always intelligible, rendering him an enigma, hard to apprehend. In fact, the biblical God seems bent on being unknowable: When asked who He is, He answers, uninformatively, that "Ehyeh-Asher-Ehyeh" ["I Am That I Am"] (Exod. 3:14).[4] Indeed, many theologians take the existence of God to be an article of faith, not subject to rational explanation.

---

4. This quotation and all subsequent quotations from the first five books of the Hebrew Bible (Old Testament) are from *The Torah: The Five Books of Moses* (1967). Quotations from other books in the Hebrew Bible are taken from *The Prophets* (1978) and *The Writings* (1982). The passage quoted is sometimes translated as "I Will Be What I Become," making

This is not the conclusion that Pascal reaches in his famous wager, which is the starting point of this book. Pascal's wager, however, is a decision—not a game of the kind that human characters play with God and that I analyze in most of this book. Because SB's superiority is undecidable in many games, his existence as well as his nonexistence does not seem demonstrable, either theoretically or empirically. This suggests that we should probably not expend great intellectual effort trying to prove otherwise but, instead, should accept undecidability as an unprovable fact of life.

Acknowledgments

I gratefully acknowledge the support of the John Templeton Foundation and discussions with my co-principal investigator, Christina Pawlowitsch. I thank Daniel Brik Laufer, Piper Devon Quinn, and Ary Reich for excellent technical assistance. People who have given me very helpful feedback are Larry Amsel, Frank C. Zagare, and two anonymous referees. My editor at the MIT Press, Emily Taber, expertly guided publication through different stages. I owe special thanks to James Bradley, the director of Templeton's project on "Randomness and Divine Providence," who provided help and encouragement throughout the project.

God's identity as becoming—that is, as the result of His decisions. That translation suggests that God is not deliberately inscrutable but rather wants to be interpreted and judged by His actions as they unfold, even if they do so in mysterious ways. As we will see, some biblical characters have faith in Him and trust in His judgment, but others do not.

1 Introduction

1.1 Background

The Bible is a sacred document to millions of people.[1] It expresses supernatural elements of faith that do not admit of any natural explanations. At the same time, some of the great narratives in the Bible do not seem implausible reconstructions of events. Indeed, biblical characters exhibit common human failings in their behavior toward one another as well as toward God.

Is it possible to reconcile the natural and supernatural elements in the Bible? This is not an easy task because God, in some concrete or ineffable form, makes His commanding presence felt in practically all biblical stories. A naturalistic interpretation of the Bible immediately confronts His uniqueness (why is there only one god and not several?) and His preternatural abilities.

Looking beyond the Bible, I attempt in this book to model a superior being who, like God in the Bible, has goals he would like to accomplish, as does a person who is in a game with SB. However, SB's supernatural properties, including omniscience and omnipotence, are not absolute, so he is limited in what he can do and induce others to do.[2]

---

1. This chapter draws from the introduction of Brams ([1980] 2003), in which I discuss the perspective that game theory casts on Bible stories and my interpretation of them. But the heart of the present book, as indicated in the preface, is not the Bible itself but several decisions and games—to which I give names—that a person (P) plays with a superior being (SB). I use these stories to raise theological, philosophical, and ethical questions about the existence of SB and what role, if any, he might play in our lives.

2. In an earlier work (Brams [1983] 2007), which did not make full use of theory of moves (TOM), I included the properties of immortality and incomprehensibility as defining features of SB. While I mention these properties later, I do not make them central to my argument that SB is undecidable. Instead, I argue that SB is undecidable because at least some of his actions appear inconsistent or otherwise inexplicable, making him a mysterious entity. Interestingly enough, God declares on more than one occasion that He is

In the Bible, the reason is clear. Although God can perform miracles and endow others with great powers, human beings, represented by P, have free will and can exercise it, even when it invokes God's wrath. (Reasons why God chose not to make human beings puppets are given later in this chapter.) Like God, SB may be thwarted in his desires.

If human beings have free will and God does not always get His way, it seems proper to model SB as a participant, or *player*, in a game. This is so because a "game"—as the term is used in game theory—is an interdependent decision situation whose outcome depends on the choices of *all* players. The difference between a decision made by P and a game played between P and SB will be discussed in chapters 2 and 3.

In the Bible, God is often frustrated by His inability to implement the outcome He wishes. When His desires are deflected or undermined by other players' choices, the Bible tells us that He becomes angry, jealous, or vengeful. But more than simply expressing these emotions, His actions are driven by them. Seen in this light, God is a very human character, despite His awe-inspiring powers.

This is not, of course, the way most monotheistic religions view God. Indeed, the Bible continually portrays Him as not only fantastically powerful but also unknowable, utterly beyond our comprehension. The implication is that He is incapable of being calculating or conniving, or stooping to the level of "playing games."

But this is not the impression that many of God's actions convey. In fact, because God often provides explicit reasons for acting in a particular way, it is hard to maintain that His motivations and desires are unfathomable.

Is it not sensible, then, to imagine SB as a game player who chooses among different courses of action to try to achieve his goals? Similarly, is it not reasonable that ordinary characters like P, knowing of SB's presence and powers, make choices to further their own ends in light of their possible consequences, including intervention by SB?

Extrapolating from the games that God and a varying cast of human characters played in the Bible, I show in this book what choices game theory and the theory of moves (TOM) prescribe as *rational* in a variety of theological games. That is, given SB's and P's preferences and their knowledge, sometimes incomplete, of each other's preferences, I

incomprehensible (e.g., by refusing to utter his name, as mentioned in the preface); at the same time, He suggests that He is immortal—exists "for all eternity" (Exod. 3:15)— implying that there was never a time when He did not exist or will not exist.

analyze the strategy choices that rational players would make to try to achieve their preferred outcomes. More precise definitions of rationality, tailored to particular circumstances, will be given later.

The games I assume that SB and P play echo some of the specific conflicts and predicaments that biblical characters faced. But more than being an echo, they enable us to ask and try to answer how SB, in exercising his superior powers, would act differently from P.

More specifically, in what ways are SB's actions distinguishable from P's? Insofar as they are not, SB's presence, and even existence, becomes harder to understand—and in certain Bible stories morally justify—thereby raising questions about his decidability.

A player is *decidable* if his/her choices indicate whether he/she is superior because he/she (i) possesses supernatural qualities of omniscience (specifically, being able to predict a player's choice before he/she makes it) or omnipotence (specifically, possessing moving, order, or threat power, to be defined later) or (ii) his/her behavior seems to occupy a higher moral plane than P's. But when SB acts no differently from what we would expect if he were P, then his identity is not decidable.

Assume, for example, that P exhibits moving power because, over generations, she is replaced by her descendants who continue to move in a game as she would. When P's behavior mimics SB's in this manner, P and SB become indistinguishable. In such a situation, we say that the identity of the player, based on moving power, is undecidable.

In other situations, we may question SB's morality when mayhem occurs that he does not stop. Worse, when SB sanctions crimes like murder, and sometimes even encourages them, his ethical superiority is cast in doubt. After considering SB's reasons for making the choices he does, I do not shy away from offering moral judgments, especially when SB's behavior is not consistent with that of a player we would consider ethically superior.

1.2   Game Theory

With the publication of *Theory of Games and Economic Behavior* (von Neumann and Morgenstern ([1944] 1953), the mathematical theory of games was born and has since been the stimulus for hundreds of books and thousands of articles on game theory and its applications. Its importance has been recognized by the award of several Nobel Prizes in economics to game theorists.

Rarely, however, has game theory been applied to humanistic material, including history and literature, although a handful of exceptions include books by Brams ([1980] 2003; 2011), Muzzio (1982), Zagare (2011), and Chwe (2014). Instead, the principal applications of game theory have been to the social sciences, mainly economics and political science, and to evolutionary biology for the analysis of competition among species.

It is true, of course, that the language of "games" and "play" has been loosely invoked in studies of the Bible. But the insights of such treatments have been just that—insights in search of a theory. Good ideas or insights, while necessary, are not sufficient for bringing coherence to a work as complex and profound as the Bible.

Because it is stories in the Bible that inspired most of the theological games that I analyze in this book, let me say something about the stories I found most amenable to game-theoretic analysis. They generally involve conflict and intrigue, wherein characters can reasonably be assumed to think about the consequences of alternative actions they might take before choosing one of them. Indeed, many of these stories serve the didactic purpose of trumpeting the virtues of cleverness and sagacity in ticklish or harrowing situations.

Contrary to the popular notion of a game, choices in game theory are not assumed to be frivolous. Quite the contrary: Players in games are assumed to think carefully about their choices and the possible choices of other players. Whether the outcome of a game is comic or tragic, fun or serious, fair or unfair, it depends on individual *choices*.

I italicize "choices," because I use game theory to explain the decisions made and actions taken in different kinds of conflictual situations, based on the preferences of the players. Thus, preferences are used to explain choices; if preferences are not entirely clear, I consider alternative rankings and determine what consequences they have for the making of rational choices. In addition, I take into account the powers that SB might exercise in influencing the outcomes of games.

Game theory and TOM are well suited for elucidating the decision-making situations I analyze, such as whether P should believe in SB, or whether SB should punish P. Because the application of these tools often requires the careful unraveling of a tangle of character motives, it imposes a discipline on the study of a situation that is usually lacking in more traditional philosophical-theological analyses.

A word about the parts of game theory I have utilized in this book is in order. I have relied primarily upon what is called the "noncooperative" theory—in which players cannot make binding or enforceable agreements—as opposed to the "cooperative" theory, in which they can. In applying the noncooperative theory, I make use of both the "extensive" (game tree) and "normal" or "strategic" (payoff matrix) forms of games in the analysis. Also, I have generally assumed only ordinal preferences (players can rank, but not attach numerical values or cardinal utilities to outcomes), and no probabilistic calculations—except, on occasion, to discuss expected payoffs, and the possible use of "mixed" strategies, when cardinal utilities can be assigned to ordinal payoffs.

I have eschewed cooperative game theory, which deals with how payoffs are distributed to players to ensure fairness, stability, or other properties of outcomes that are considered desirable. For one thing, most players have difficulty assigning utilities to outcomes, but they can more readily rank them from best to worst.

This is certainly true of the Bible stories, for which there is generally insufficient information to quantify payoffs and make expected-payoff calculations, whereas it is possible to order outcomes by ranking them from best to worst. Also, cooperative game theory assumes that contracts can be written that are binding or enforceable, whereas noncooperative game theory does not make this assumption but instead asks what strategies lead to equilibrium outcomes.

If the more esoteric and mathematically deeper aspects of game theory offer little help in interpreting the theological games between SB and P that I analyze, it is fair to ask whether the elementary, nonquantitative theory has the depth and power to cast new and penetrating light on these games. I believe TOM helps in this regard by assuming that the players are not myopic decision makers but think ahead and, in addition, may have different capabilities that affect the choices they make. I try in all cases to link the more formal analysis of games with a verbal explication of the strategic situations that SB and P face to lend plausibility to the explanation, based on the game played, of the outcomes they chose.

My analysis of the games and the rational explication of player choices in them will in some cases be controversial. When the reader disagrees with assumptions I have made about the players, their strategies, the possible outcomes, and the players' ordinal preferences (i.e., rankings) of the outcomes, I urge him or her to experiment with different assumptions. The game-theoretic framework should not be rejected out of hand simply because there are alternative—if not superior—representations of the strategic situations that I present.

Equally controversial, if not shocking, may be the philosophical, religious, and theological implications I have drawn from the analysis.

I have not shunned these, because I believe that the ultimate significance of this analysis rests on the ramifications it has for understanding
SB and his relationship to the human world he did (or did not) create.
Because SB's place in the scheme of things helps to define P's place as
well, I believe the games I analyze also have something important to
say about the human experience and its spiritual connection.

Indeed, this connection is specifically defined by the games that SB
and P play. Only a detailed and searching study of the strategies players
choose and the outcomes they obtain provides a touchstone for com-
paring and understanding the significance of one's own choices and
experiences in life. This certainly is one of the benefits of looking back
at stories in the Bible and then forward to their possible consequences
for the choices we make today, as well as our place in the universe.

1.3   The Question of Free Will

To commence the analysis, it is reasonable to ask why God created the
human world from one that was "unformed and void" (Gen. 1:21) in
the first place. The Bible says simply that God acted; there were no
other players, so He was not playing a game with anybody else. Leszek
Kolakowski thinks, with a touch of irreverence, that the reason He
created the world is evident:

God created the world for His own glory. This is an indisputable fact and one,
moreover, that is quite understandable. A greatness that nobody can see is
bound to feel ill at ease. Actually, under such circumstances one has no desire
whatsoever to be great. Greatness would be pointless, it would serve no
purpose. ... Holiness and greatness are possible only in a concrete setting. ...
And only then [after the creation of the world] did He really become great, for
now He had someone who could admire Him and to whom He could compare
Himself—and how favorably! (Kolakowski 1972, 3)

Was it loneliness, and the need to be admired, that drove God to
create the world? There is perhaps a hint of what God sought in the
biblical statement that there was "no man to till the soil" (Gen. 1:27).
But God did not have to create man "in His image" (Gen. 1:27) to
accomplish such a mundane agricultural task. God obviously had
better things in mind for men and women when He directed them to

be fertile and increase, fill the earth and master it; and rule the fish of the sea;
the birds of the sky, and all the living things that creep on earth. (Gen. 1:28)

If this were not enough,

The LORD God planted a garden in Eden, in the east, and placed there the man whom He had formed. And from the ground the LORD God caused to grow every tree that was pleasing to the sight and good for food, with the tree of life in the middle of the garden, and the tree of knowledge of good and bad. (Gen. 2:8–9)

But the garden of Eden was not quite idyllic, nor its human inhabitants quite so pure of heart, that conflict did not arise. In fact, God set the stage for a challenge to His authority when He commanded man:

Of every tree of the garden you are free to eat; but as for the tree of knowledge of good and bad, you must not eat of it; for as soon as you eat of it, you shall die. (Gen. 2:16–17)

This admonition would hardly be necessary if man (and later woman) were simply God's puppets, blindly attentive to His every wish and command.

God took six days to create the world, proclaiming at the end of each of the first five days that the results were "good"; at the end of the sixth day, after creating man and woman and other living creatures, God found it "very good" (Gen. 1:31). On the seventh day, He blessed this day and called it holy.

Manifestly, God was a good planner, judging the fruits of His labor to be acceptable at each stage before moving on to a new stage. It therefore seems unlikely that He would have created a person with free will whom He could not control if this were not His intention. Thus, in creating the world, I assume God most preferred to create someone with free will; failing that, I presume a person as a puppet would be better than no person at all. After all, if God truly craved admiration, as Kolakowski maintains, it would be better to be admired by a person created in one's own image than by ordinary animals.

Worse, I suppose, for God would be not to create a person at all and have only the plants and animals to behold. Moreover, without people, there would be nobody to rule over other creatures. Still worse would be to remain a solitary God without even a world to look down upon, much less anyone to proclaim His glory. In summary, I assume God's ranking of alternatives from best to worst was the following:

1. Create a person with free will.
2. Create a person as a puppet.
3. Create a world without people.
4. Create no world.

By choosing His most-preferred alternative, God acted rationally.

God's preference for populating the world with men and women who possess free will, rather than just creating puppets He could completely control, requires further justification, especially in light of the grief and anguish that human beings later caused Him. (If God were omniscient, could not this be anticipated?) To begin with, I believe that as much as God sought glory and desired praise, He recognized that it would be hollow if men and women were but His puppets.

Put another way, God wanted to receive homage that was heartfelt, not forced or dictated. As Elie Wiesel argued, "God loves man to be clear-sighted and outspoken, not blindly obsequious" (Wiesel 1977, 104). This view is supported by evidence presented in later chapters that God most relished unswerving faith when people, especially His chosen people (the Israelites), were in the direst straits. In such circumstances, the faith of persons with free will, not puppets, can be tested.

God relentlessly tests men and women throughout the Bible, giving them multiple opportunities to sin. The results are mixed: Sometimes they succeed impressively; sometimes they fail miserably; and sometimes they falter before regaining their faith. If they were unalloyed successes in all these tests, then the world would be a dull place— precisely, it seems, what God wanted to escape from by creating the world in the first place.

To make the world less predictable and therefore more alive and engrossing, it was in God's interest to give men and women free will— or at least hold them responsible for their actions even if they are not all freely determined (Pharaoh is a case in point, to be discussed in section 9.3). The price He paid, of course, is having to contend with people who continually frustrate Him and occasionally drive Him to the brink of despair. But when men and women succeed, God can not only feel proud but also rest assured that their profession of faith is genuine and not more mere flattery, which presumably makes their character defects worth tolerating.

Not surprisingly, God is selective in what kinds of men and women He supports. As befits a demanding creator, He continually makes judgments about the deeds and misdeeds of His subjects. Having made up His mind, He is not above playing favorites, as we will see in many of the stories.

2 Belief Decisions

2.1 Introduction

Theology studies the relationship between human beings and God.[1] Game theory seems well suited for analyzing this relationship, though some may challenge the seemingly impious view that God plays games with us.[2] But this view is inherent in those Western religions that presume God can be conceived of in personal terms and, more specifically, is in a one-to-one relationship with each of us.

The idea of a personal God, especially one who is a game player, is alien to most Eastern religions. I confess that the present analysis has a Western religious bias, perhaps epitomized by the view of Martin Buber (1958, 135): "The description of God as a Person is indispensable for everyone who like myself means by 'God' not a principle ... not an idea ... but who rather means by 'God,' as I do, him who—whatever else he may be—enters into a direct relation with us."

Despite this focus on a personal God in a direct relation with us, however, I consider in this chapter the possibility that God may more accurately be conceived of as a "state of nature"—not a player in a game as such—with nature being neutral or indifferent. This contrasts with the view of God that I offer in chapter 3 and later, in which God is in a two-person game with human players.

In this chapter I do not, at the outset, presuppose the existence of God. Instead, I ask in section 2.2 a theological question concerning the relationship between a person (P) and a superior being (SB): Is it

---

1. This chapter is adapted, with modifications, from Brams 2011, chap. 3.
2. This will be true for those who see God as a lofty and majestic figure, someone of overpowering grandeur and infinite wisdom and strength. But this view is challenged by the picture of God drawn in the Hebrew Bible, in which He is variously angry, jealous, petulant, or arbitrary (Brams [1980] 2003).

rational for P to believe in SB's existence? While SB may be thought of as God, I assume SB may be some other religious figure (e.g., Jesus Christ), or a secular force with supernatural powers.

The rationality of belief in God in the Judeo-Christian tradition is a venerable question that has been the subject of an enormous body of literature, both classical and modern. The most famous rational argument for believing in God is that of the seventeenth-century French polymath Blaise Pascal, whose wager to justify such belief I analyze in section 2.2. Later in this section, I reframe P's choice in the Search Decision, which incorporates the possibility that a search by P may be indeterminate.

Pascal's approach is decision-theoretic—he does not assume we play games with God, in which God actively chooses strategies. Instead, he supposes that each person makes a calculation about whether or not believing in God in an uncertain world is justified (Landsberg 1971; Rescher 1985; Chimenti 1990; Jordan 2006).

This kind of choice is a decision in a one-person game, or a *game against nature*, wherein "nature" is God or some other SB. Unlike a player in a two-person game, God, if He exists, is assumed to be neither benevolent nor malevolent, though I do suggest, in the case of the Search Decision, that Pascal's formulation of the decision to believe or not believe becomes more like a game if God or SB can influence the choice of a state of nature.

I do not analyze the so-called proofs of God's existence (or nonexistence) in this chapter. In an excellent synthesis of the literature, beginning with Descartes, on the existence of God, Hans Küng (1980) argues that all such proofs are flawed, and I share his view. In place of proof, Küng argues for a "rationally justified" faith, which of course does not have the force of a logical proof but, as the subtitle of his book suggests, is "an answer for today." This answer, according to Küng, is rooted in developing a fundamental "trust in reality"; Pascal's wager may be seen as a reason, perhaps cynical, to develop this trust.

In section 2.3, I introduce the Concern Decision, wherein P chooses whether or not to be concerned with SB, which in turn depends on whether or not SB is aware and cares about P. I show that, as in the Search Decision, P does not have a dominant strategy (to be defined and illustrated shortly). Rather, her best choice depends on the state of nature, though SB may influence P's choice if he can signal whether or not he cares.

The question of whether this answer is satisfactory, as a religious response, I leave to theologians and philosophers of religion. Instead,

I focus here on a related, if more subjective, question: Is it rational to *believe* in SB's (or God's) existence—and, thereby, search for evidence or even be concerned? I stress that the rationality of theistic belief is separate from its truth—a belief need not be true or even verifiable to be rational if, for example, it satisfies certain psychological or emotional needs of a person. Here, however, I tie the rationality of belief to the "evidence" at hand. This theme will be developed further in chapter 3, in which I assume that SB is not a state of nature but rather an active player in a game with P.

It is worth noting that the question of God's existence is almost never raised in the Bible. When it is, as in Moses's confrontation with Pharaoh in the Book of Exodus (see section 9.3), it is the tangible evidence of God's miraculous powers that settles the issue for Pharaoh, at least temporarily. For many today, however, the evidence is not so compelling.

2.2 Pascal's Wager and the Search Decision

In his *Pensées*, published posthumously in the late seventeenth century, Pascal ([1670] 1950) assumes a person is in a betting situation and must stake her destiny upon some view of the world. Starting from the agnostic assumption stated in no. 223 that "if there is a God ... we are incapable of knowing what He is, or whether He is," and "reason can settle nothing here ... a game [!] is on," Pascal proceeds to invoke reason to say that a prudent person, in her cosmic ignorance, should bet her life on God's existence (for Pascal, this meant in the Roman Catholic faith). If one believes, then the two possible states of nature that may occur have the following consequences:

1. God exists: One enjoys an eternity of bliss (infinite reward).
2. God does not exist: One's belief is unjustified ("loss of nought"), or, at worst, one is chagrined for being fooled (finite penalty).

Because state 1 promises an infinite reward if one believes, and leads at worst to a finite penalty if God does not exist, the choice seems clear to Pascal: One should believe in God's existence.

This argument, made in no. 223 of *Pensées*, implies that belief in God is justified by the infinite reward and only finite penalty, but it needs elaboration. A person, P, is choosing between belief and nonbelief, as shown by the two rows of the outcome and payoff matrix in figure 2.1 (the payoffs are ranks, ranging from best, or 4, to worst, or 1).

State of nature

	God exists	God doesn't exist
Believe in God: B	Belief justified: infinite reward (4)	Belief unjustified: finite penalty (2)
Don't believe in God: $\overline{\text{B}}$	Nonbelief unjustified: infinite penalty (1)	Nonbelief justified: finite reward (3)

P

Key: 4 = best, 3 = next-best, 2 = next-worst, 1 = worst for P

Figure 2.1
Outcome and payoff matrix of Pascal's wager

In this situation, it is proper to compare the "expected payoffs" that each of these alternative courses of action yields and choose the higher one. Pascal is not explicit about this, but the calculation of expected payoff for nonbelief that I next describe is suggested by his discussion of the "fear of hell" in no. 227 of *Pensées*.

Pascal's argument, as stated earlier, holds even if it is not true, as Pascal assumed, that "the chances of gain [if God exists] and loss [if God does not exist] are equal." For however small the chance of state 1 (God exists) is, when multiplied by an infinite gain, the resulting *expected payoff* (sum of utilities of outcomes times their probabilities of occurrence) for believing in God's existence is infinite. Because an infinite payoff has no obvious meaning, one might think of this payoff as some stupendous, but finite, reward.

By comparison, if one bet that state 2 (God does not exist) was true, but it turned out to be false, one would suffer an eternity of torment (infinite penalty), or a huge loss from not believing. (Henceforth, stupendous rewards and huge penalties might be substituted for infinite gains and infinite losses to make this discussion more concrete—and mathematically acceptable since, technically, one would evaluate a player's expected utility as either the reward or penalty approaches, in the limit, positive or negative infinity.)

To summarize, believing in God's existence yields an infinite positive expected payoff (infinite reward minus finite penalty), whereas not believing yields an infinite negative expected payoff (infinite penalty minus finite reward). Because this calculation can be made before the outcome is known (if it ever is!), it suggests that P's key to happiness—and perhaps heaven—is being prepared for the possibility that God exists by believing in Him. One should bet that He does by accepting the wager.

Even if one should not attain eternal bliss, at least it can be said that one made an honest effort to achieve it. Furthermore, Pascal avers that the very act of believing in this calculated fashion sets up the conditions for developing genuine faith and becoming a true believer, so there is nothing insincere or dishonest about starting off by making the expected-payoff calculation, though one's faith is ultimately sustained "by taking holy water, by hearing mass, etc."—the accoutrements of religion.

One weakness, I believe, in Pascal's argument is that he never postulated a third possibility (or still others): SB's existence is *indeterminate*, because information that would settle this question is unattainable. I next add this as a state of nature, or a situation that might arise in the world.

Now the three states of nature, which I assume P perceives, are the following in what I call the Search Decision:

1. SB's existence is verifiable.
2. SB's nonexistence is verifiable.
3. SB's existence or nonexistence is indeterminate (insufficient information).

True, state 3 may in fact hide one of the other two states (e.g., because the search did not go on long enough), but P may simply be incapable of verifying SB's existence or nonexistence. If the latter is the case, state 3 and the other two states are mutually exclusive in P's eyes: Observing or experiencing state 3 precludes observing or experiencing states 1 and 2, just as the latter two states preclude each other and state 3.

It is perhaps strange that Pascal did not postulate state 3 in his wager, because he said that God, being "infinitely incomprehensible," forecloses the possibility that His existence can ever be determined. Pascal's wager, in other words, presumes a bet whose outcome will never be known—at least in one's present life—so one must be wagering about one's rewards or penalties in an afterlife.

To put it somewhat differently, Pascal seems to have thought we will experience only indeterminacy in our present lives. Nevertheless, he postulates the possibility of only the determinate states, presumably because he thinks we will learn of God's existence or nonexistence in the hereafter. I am not so sure this is the case and, therefore, prefer to retain state 3 in the calculations we make while alive.

The three postulated states of nature in the Search Decision are shown in figure 2.2. Combined with P's two strategies of searching (S) and not searching ($\bar{S}$) for SB, there are six possible outcomes. I interpret S to mean that P tries to learn about SB's existence or nonexistence, whereas $\bar{S}$ means P makes no effort to solve the great mystery, if you will, about SB's existence.

P's ranking of payoffs associated with the six possible outcomes are also shown in figure 2.2, with 6 being the highest and 1 the lowest. I assume that P would most like the search to verify SB's existence (6) and least like it to fail (1). In between, P would prefer to search even if the search only succeeds in verifying SB's nonexistence (5), or not search if SB's existence is indeterminate (4).

Worse than the latter outcome is when searching would have uncovered valuable information—in particular, information that would verify SB's nonexistence (3) or existence (2)—but P, by not searching, makes no effort to find out which of these states occurs. Conceivably, P might rank the last outcome worst—switching 2 and 1 in figure 2.2—because it would be more damaging to ignore available information on SB's existence than to search and find nothing. This alternative ranking of P's two worst outcomes, however, does not affect the argument I make in the next paragraph.

Whatever utilities, which I assume are finite, one attaches to the outcomes consistent with the ranks in figure 2.2, P does not have a

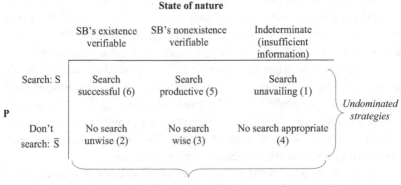

Undominated strategies (if SB is a player with the same preferences as P)

Key: 6 = best; 5 = next-best; . . . , 1 = worst for P

Figure 2.2
Outcome and payoff matrix of Search Decision

dominant, or unconditionally best, strategy: S is better than $\bar{S}$ if SB's existence or nonexistence is verifiable (states of nature 1 and 2); but $\bar{S}$ is better than S if the situation is indeterminate (state of nature 3). This complicates P's decision, because her rational choice depends on the (unknown) state, which renders her strategies of believing or not believing *undominated*—one is not always better than the other.

To be sure, if state of nature 1 occurs, P realizes an infinite reward (eternal bliss) from choosing S and an infinite penalty (eternal torment) from choosing $\bar{S}$. Clearly, these infinite values swamp the positive or negative values of other outcomes, making P's choice clear: She should choose S and believe if state 1 occurs. But if these high and low values are not infinite, the expected payoff for choosing S will not necessarily exceed that for choosing $\bar{S}$, especially if the probability of state 3 is very high. Without having this information about states of nature, however, P will remain in a quandary, able only to ponder the imponderable. (This quandary also holds in Pascal's wager, because neither belief nor nonbelief is a dominant strategy—as in the Search Decision, P's strategies are undominated—but postulating infinite rewards and penalties skewed things in favor of belief.)

For the sake of argument, assume that the states of nature do *not* occur by chance, each with some nonzero probability, but that SB can choose which one will arise, or can induce P to think that one has occurred.[3] That is, SB can choose a state, like a strategy in a game, that would indicate that (1) his existence is verifiable, (2) his nonexistence is verifiable, or (3) his existence cannot be determined. As an illustration, SB might indicate state (1) by signs of revelation, or state (3) by giving no signs. (How SB would indicate his nonexistence is verifiable in state 2 is not apparent.)

What state will SB choose? Like P, SB has no dominant strategy in this decision-transformed-into-a-game if SB's ordering of the six outcomes is exactly the same as P's. But in such a game of total agreement, a mutually best (6,6) outcome would almost surely induce both players to choose their strategies associated with it.

However, as I will argue in the case of the Belief Games in chapter 3, SB has good reason *not* to make his existence apparent, though he

---

3. In decision theory, this is a violation of what Binmore (2009, 5, 31) calls "Aesop's principle" (based on one of Aesop's fables), whereby the states of nature are assumed independent of a player's choices. In the Hebrew Bible, God frequently violates this principle, inducing states that He thinks will influence a character's choices. In chapter 3, I explicitly model these interdependent choices between SB and P in two full-fledged games.

might want to signal that state 1, and maybe state 2, should not be ruled out in order to push P in the direction of searching. This may induce P to be obedient, too, though to what end is not clear.

To most agnostics, the signals, if they hear or see any, are ambiguous. This probably predisposes them to think indeterminate state 3 is most probable, and hence their search should not continue, unless they associate very high utility with being successful in their search (in improbable states 1 and 2).

I will not speculate further on P's choices in this situation—or on SB's choices if in fact he has them in a game wherein he can choose a state of nature. I note, however, that P's apparent certitude in favor of belief in Pascal's wager may be dashed in the Search Decision, because P does not have a dominant strategy in the absence of infinite rewards and penalties. Moreover, an expected-payoff calculation that favors S or $\bar{S}$ cannot be made without information on the probabilities of the states. If, following Pascal's recommendation, P chooses to believe anyway and so commences a search, she will be severely disappointed if state 3 turns up, giving P's worst (1) outcome.

The lack of a clear-cut choice in the Search Decision is likely to be unsatisfactory to those who seek unambiguous answers, not qualifications. These qualifications, however, cannot easily be dismissed in any rational assessment, though one might contend that belief in SB (or God) is not, or cannot be, a rational choice. William James ([1902] 1967, 723), for example, maintained that our beliefs ought to be determined by our "passional nature" when their truth cannot "be decided on intellectual grounds."

But "passions," like preferences or goals, are neither rational nor irrational—it is the choices based on them that are. Rational choices are those that best achieve one's goals. Although passions tend to be more identified with emotions than logic, to be "driven" by one's passions is in fact tantamount to trying to find the best means to attain an end, as we will see in later chapters (for more on the role that passions or emotions play in rational decision making, see Brams 2011, chap. 7).

Even if one cannot articulate a logic to one's passions, is it not rational to try to satisfy them? I explore this question by analyzing a different decision in the next section.

2.3 The Concern Decision

In the Concern Decision shown in figure 2.3, P's choices are to be concerned or not with SB; the states of nature for SB are that he is aware,

State of nature

	SB aware and cares whether P is concerned	SB not aware and does not care whether P is concerned
Be concerned: C	Concern justified (4)	Concern unjustified (2)
Be unconcerned: $\overline{C}$	Unconcern unjustified (1)	Unconcern justified (3)

P

Undominated strategies

Key: 4 = best; 3 = next-best; 2 = next-worst; 1 = worst

Figure 2.3
Outcome and payoff matrix of Concern Decision

and so cares about, what P chooses, or is not aware and does not care. I assume that SB's awareness implies that he cares, and his unawareness implies that he does not care, so awareness and caring by SB are inextricably linked. Unlike the Search Decision, the Concern Decision imputes feelings of awareness and caring to SB, and by extension concern by P about the consequences of her choices. We ask: Is P's concern or unconcern justified or not?

Whether P's choice to be concerned (C) or not ($\overline{C}$) is viewed as a logical decision or an emotional response, the rational calculus is the same. Of course, passions may dictate a different assignment of utilities to outcomes than "mere" preferences, but this has no bearing on P's making a rational choice that mirrors her passions or preferences as long as they are consistent with P's ranking of the outcomes that I have assumed.

If SB does not care, I assume that P is better off choosing $\overline{C}$ and not expending any effort (even if P is empathetic, because SB in this case is not worthy of empathy). I make no assumption about SB's preferences, only that he may or may not be aware and care.

P most prefers C when SB cares (4, justified concern); next best for P is $\overline{C}$ when SB does not care (3, justified unconcern). These outcomes rank higher than P's choosing C and SB's not caring (2, concern unjustified) and P's choosing $\overline{C}$ even though SB cares (1, unconcern unjustified). The latter outcome is P's worst, because she abandons an SB who cares; in Pascal's wager, this would lead to unmitigated agony for P.

Even though P's choice here—to be concerned or not about SB—is not exactly the way Pascal presented alternatives in his wager, Pascal, I imagine, would advise as follows: Be concerned, for there is an infinite reward associated with rank 4, and an infinite penalty with rank 1, whereas the payoffs associated with ranks 2 and 3 are finite and so are not consequential in an expected-payoff calculation. More precisely, whatever the probabilities associated with each state of nature are, as long as they are positive, there is an infinite gain associated with being concerned, and an infinite loss with not being concerned.

If one does not assume infinite rewards and penalties, however, but considers only P's ranking of the four possible outcomes, P would be in a dilemma. Without a dominant strategy, her best choice depends on what state of nature arises.[4]

Most theists, I presume, would assert that God is aware and cares, whereas most atheists would say that the question is meaningless, because God does not exist. Thus, because of their different "passional natures"—believers with a passion for a caring God, nonbelievers with a passion against believing in God—neither would have a problem about which choice to make in the Concern Decision. For agnostics, on the other hand, being concerned or unconcerned are undominated strategies, based on P's assumed ranking of the four outcomes.

If this was indeed Pascal's ranking, he could, as I suggested earlier, resolve the dilemma by assigning infinite positive utility to rank 4 and infinite negative utility to rank 1. Practically speaking, however, what do these colossal rewards and penalties signify?

They might be considered a gauge of P's very strong passions, but are they meaningful if P thinks that the probability of the first state (SB aware and cares) is negligible? If, instead, SB is much more likely to be unaware or indifferent, perhaps these attributions of infinite value should be ignored in favor of choosing between the middle rankings in the second state (SB not aware and does not care). In this case, P's decision would favor $\bar{C}$ (because 3 > 2).

One possible way around the dilemma created by undominated strategies would be to alter P's ranks, switching, say, 4 and 2 or 4 and 1. In the former case, C would be a dominant strategy; in the latter case, $\bar{C}$ would be a dominant strategy.

---

4. This question is also posed by another conundrum, Newcomb's problem, which I discuss in chapter 5 and which is intimately tied to the game of Prisoners' Dilemma (Brams 1975; [1983] 2007); see also Lewis 1985 and Hurly 1994.

However, I find it difficult to justify these new assignments, which say that P's best outcome is to be concerned when SB does not care, or to be unconcerned when SB does, reversing the preferences it seems reasonable to impute to P. These switches smack of playing with numbers to sidestep a genuine dilemma; in my opinion, they are quite absurd.

What may be more defensible is to switch 3 and 4, arguing that justified unconcern, which might save P much time and effort, is better than justified concern. After all, each expression—of concern or unconcern—is consistent with the "facts" (i.e., the state of nature). Similarly, switching 1 and 2, based on a similar argument that to be concerned is more demanding than being unconcerned (even unjustifiably), also seems defensible.

Unfortunately for P, neither the 3–4 nor the 1–2 switch, or even both together, would endow her with a dominant strategy. The dilemma would remain: P's preferred choice, at least based on these rankings, would depend on the state of nature that arises.

This dilemma also underlies the Search Decision. To see this, simplify this decision by removing the first state of nature (SB's existence verifiable) or the second state (SB's nonexistence verifiable) in figure 2.2. Now, with either state removed, P's two best outcomes lie along the main diagonal of the payoff matrix (from upper left to lower right); and her worst two worst outcomes lie along the off-diagonal (from lower left to upper right). This placement of preferred and nonpreferred outcomes implies that P does not possess a dominant strategy. Moreover, no switching of either the diagonal or the off-diagonal ranks changes this lack of dominance.

To return to the Concern Decision, one's passions, such as those Pascal assumed for heaven and against hell, may offer relief by decisively throwing the choice one way or the other if the utilities of rewards or penalties are infinite. If not, however, the rankings by themselves that I have assumed do not, because P's strategies in both the Search and Concern Decisions, as well as Pascal's wager, are undominated.

For the agnostic who has the assumed preferences, the absence of a best choice, independent of what state of nature arises, poses a problem: It complicates the easy calculus of Pascal's wager, undergirded by infinite rewards and penalties, for which the dominant strategy is to believe. A likely consequence for P is indecision, and possible anxiety if she finds agnosticism uncomfortable rather than finding a resolution in belief (or nonbelief).

2.4   Conclusions

In the Search Decision and Concern Decision, P's choice of searching for, or being concerned about, SB depends on the verifiability of SB's existence, or on SB's awareness and caring for P. Pascal's neat solution to this intellectual puzzle was to suppose payoffs for P that postulated a halcyon heaven and a horrible hell. Thereby Pascal was able to justify belief in God—or at least P's appearing to believe—as infinitely rewarding, even if this belief turned out to be false.

Pascal's payoff assumptions may strike agnostics as loading the dice to prop up a result that Pascal sought to support with rational arguments. By contrast, a rational choice in the Search Decision and the Concern Decision is not so clear-cut, because P does not have a dominant strategy in either without Pascal's reward and penalty assumptions. Instead, her strategies are undominated, whereby her best choice depends on the state of nature.

Our "passional nature," in James's felicitous phrase, may help resolve this difficulty if it implies utilities and probabilities that swing the balance toward the choice of one strategy. While goals are neither rational nor irrational, the utilities we assign to achieving them, when multiplied by their probabilities of occurrence and summed, yield expected utilities that, except for ties, render one strategy more beneficial than others.

But in the Search Decision, an agnostic P is likely to attribute a high probability to SB being indeterminate, so his existence (or nonexistence) is not likely to be ascertainable. If this is true, P's rational decision would be *not* to search. Similarly, in the Concern Decision, if P believes that it is likely that SB is not aware and does not care whether P is concerned, she would be justified in *not* being concerned herself.

I would add that SB's indeterminateness, or unawareness and lack of care, are not sufficient grounds for saying that SB does not exist. Instead, he may just be unknowable, and so one should, as an agnostic, keep an open mind—even if there is no satisfactory resolution in the end—because SB's existence is indeed undecidable.

3 Belief Games

3.1 Introduction

In this chapter, I construct two two-person Belief Games, which I will refer to as Belief Game 1 and Belief Game 2, in which P and SB each can make choices. In these games, SB, rather than just being a passive state of nature as assumed in chapter 2, is not indifferent but instead, like P, has preferences and makes strategy choices that affect the outcome.

In particular, SB must decide whether to reveal himself or not. Revelation provides evidence for P to believe in SB's existence, but it is something that SB would prefer not to provide, because it does not test P's faith (i.e., her belief in SB without evidence). As in chapter 2, P, independently, must decide whether to believe or not believe in SB's existence. These two choices of each player define a 2×2 game, which can lead to four possible outcomes.

Of course, to model the relationship between P and SB as such a game drastically simplifies a deep and profound religious experience for many people. My aim, however, is not to describe this experience but to abstract from it, using games to analyze a central theological question: Can belief in SB be conceptualized as a rational choice when SB is not a state of nature but has his own goals?

The answer depends, in part, on whether it is proper to view SB as a game player, capable, like P, of making independent choices. Or is SB too ethereal or metaphysical an entity to characterize in these terms? If God, as Buber put it, is in a "direct relation" with us (see section 2.1), it is not a great leap of faith, in my view, to model SB's relationship to P as a game.[1]

---

1. As Cohen (1991, 24) points out, however, in the non-Western world "the concept of a personal, unmediated relationship between a human being and deity is quite incomprehensible."

In the game, SB has two strategies: Reveal himself (R), which establishes his existence, or don't reveal himself ($\overline{\text{R}}$), which fails to establish his existence. Similarly, P has two strategies: Believe in SB's existence (B), or don't believe ($\overline{\text{B}}$).

Belief Game 1 differs from Belief Game 2 in attributing different goals to P. In Belief Game 1, P prefers to believe in SB's existence—if there is evidence that SB exists—than not believe if there is no evidence, whereas in Belief Game 2 the reverse is the case. In effect, Belief Game I reflects P's preference for belief over nonbelief when there is positive evidence, whereas Belief Game 2 reflects her preference for nonbelief over belief when evidence is lacking. I will spell out these differences more precisely in subsequent sections.

Both games turn out to be "cyclic," which roughly means that players, looking ahead, have reason to change their strategy choices to try to achieve better outcomes. I ask where these games might come to rest if one player has "moving power," which means that he or she can continue moving to try to force the other player to stop at a preferred outcome for the moving player (presumably SB, but not always).

I illustrate moving power in cyclic games by developing, and justifying in some detail, Belief Game 1. I then turn to Belief Game 2 and compare outcomes in the two games.

The analysis in this chapter is a prelude to the further development of the theory of moves (TOM) in chapter 4 and its application to Belief Games 1 and 2. I stress that the concepts and rules of play of game theory and TOM, coupled with the presumed goals of the players, are key. If other plausible goals lead to different conclusions about rational play, move dynamics, or equilibrium outcomes, they are fair game (no pun intended).

3.2   Belief Game 1

To describe the first of two games that I posit between SB and P, I begin by specifying the players' preferences for the four possible outcomes of Belief Game 1.[2] However, I do not attempt to give numerical values (or cardinal utilities) of the players for each outcome. Instead, I assume that the players, as in chapter 2, can rank the outcomes from best to worst according to the following scale:

---

2. Earlier I called this the Revelation Game when there was only one Belief Game (Brams 1994, 2011).

4 = best; 3 = next best; 2 = next worst; 1 = worst.

Hence, the higher the number, the better the outcome, which I will refer to as the *payoff* to a player.[3]

The first number, x, of each pair of payoffs (x,y) indicates the payoff to the row player (SB); the second number, y, indicates the payoff to the column player (P). Thus, for example, the payoff (2,3) indicates the next-worst payoff to SB and the next-best payoff to P.

A game in which each player knows the other player's preferences—as well as his or her own preferences—is called a *game of complete information*. When players not only possess this information but also know that they know it, and so on ad infinitum, they are said to have *common knowledge*. Henceforth, I assume games in this book are games of complete information and common knowledge unless otherwise indicated.

Instead of trying to justify the preferences of SB and P for each outcome, I begin by specifying each player's (i) primary and (ii) secondary goals:

SB: (i) wants P to believe in his existence; (ii) prefers not to reveal himself.

P: (i) wants belief (or nonbelief) in SB's existence confirmed by evidence (or lack thereof); (ii) prefers to believe in SB's existence.

The primary and secondary goals of each player, taken together, completely specify a player's ordering of outcomes in a 2 × 2 game. The primary goal distinguishes between the two best (3 and 4) and the two worst (1 and 2) outcomes of a player, whereas the secondary goal distinguishes between 4 and 3, on the one hand, and 2 and 1 on the other.[4]

Thus for SB, goal (i) establishes that he prefers outcomes in the first column (4 and 3) of the figure 3.1 outcome and payoff matrix

---

3. This is somewhat an abuse of terminology, because payoffs in standard usage are cardinal utilities, not ranks. These ranks are sometimes called "ordinal utilities," because they give an ordering of outcomes. However, they do not indicate how much better a higher-ranked outcome is than a lower-ranked outcome; moreover, for the same ranks, the players may have quite different cardinal payoffs. But the ranks are more general than cardinal utilities, because they subsume all possible cardinal payoffs that are consistent with them.

4. This is an example of a lexicographic decision rule, whereby outcomes are first ordered on the basis of a most important criterion, then a next-most important criterion, and so on (Fishburn 1974).

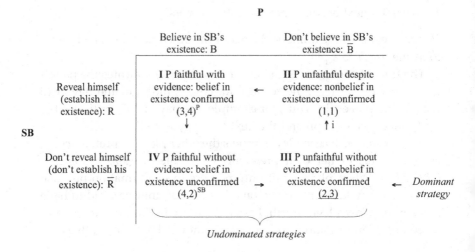

Figure 3.1
Outcome and payoff matrix of Belief Game 1 (game 48)

(associated with P's strategy of B) to outcomes in the second column (1 and 2) of the matrix (associated with P's strategy of $\bar{B}$). Between the two outcomes in each of the columns, goal (ii) establishes that SB prefers not to reveal himself (hence, 4 and 2 are associated with $\bar{R}$) over revealing himself (so 3 and 1 are associated with R). In the appendix, which gives a complete listing of all 2×2 strict ordinal games of conflict and their properties, Belief Game 1 is game 48.[5]

Likewise for P, goal (i) says that P prefers to have her belief or non-belief confirmed by evidence (so the main-diagonal outcomes from upper left to lower right are 4 and 3) to being unconfirmed (so the

---

5. *Games of conflict* are those in which the players strictly rank the four outcomes from best to worst (there are no ties), and there is no mutually best (4,4) outcome. These games are distinct in the sense that no interchange of players or their strategies can transform one game into another. There are 57 such games; if one includes the 21 games with a (4,4) outcome, there are a total of 78 distinct 2×2 strict ordinal games. All except three of the 78 are games of partial conflict (including Belief Game 1), in which the preferences of the players are not diametrically opposed; for an example of the latter, see section 6.2.

off-diagonal outcomes from upper right to lower left are 1 and 2). Between the pairs of main-diagonal and off-diagonal outcomes, goal (ii) says that P prefers belief with revelation (4) to nonbelief without revelation (3), and belief without revelation (2) to nonbelief with revelation (1).

Goal (ii) of P implies that she is happier being a believer (4) than a nonbeliever (3), given that goal (i) is satisfied—namely, that there is positive evidence for belief and negative evidence, or at least a lack of evidence, for nonbelief.[6] Shortly I will analyze the optimal strategies of SB and P in Belief Game 1 using game theory and TOM, but first I consider both the reasons for and sources of the players' goals which, I believe, lend them plausibility. Together, the players' primary and secondary goals completely define their preferences for the four outcomes.

In the contemporary world, evidence from one's observations, experiences, and reflections accumulates, predisposing one to believe or not believe in the existence of God or some other supernatural being or force—or leaves the issue open, as I suggested might be the case by postulating indeterminacy as a state of nature in the Search Decision in section 2.2. How beliefs are formed about a possible deity is less well understood.[7]

Of course, religions predispose one toward particular views, which religious works may reinforce. I next offer some brief remarks on the Hebrew Bible, which may lend plausibility to the goals of P and SB that I assume.

Evidence that the biblical God wanted His supremacy acknowledged by both Israelites and non-Israelites is plentiful in the Hebrew Bible. Moreover, the biblical narratives make plain that God pursued this goal with a vengeance not only by severely punishing those who did not adhere to His commands and precepts but also by bestowing rewards on the faithful who demonstrated their unswerving belief through good deeds or sacrifices.

Yet beyond providing indirect evidence of His presence through displays of His might and miraculous powers, the biblical God has an

---

6. As I will show later, Belief Game 2 reverses P's preference, making her happier being a nonbeliever when there is no evidence (i.e., no revelation by SB) than a believer when there is.

7. For a developmental analysis of faith, see Fowler (1981). Different kinds of theological evidence, and the different kinds of rational calculations that underlie them, are discussed in Swinburne 1981, chaps. 2 and 3.

overarching reason for not revealing Himself directly:[8] It would undermine any true test of a person's faith, which I assume to be belief in God not necessarily corroborated by evidence. Only to Moses did God confirm His existence directly—"face-to-face" (Exod. 33:11; Num. 12:6–8; Deut. 34:10)—but whether Moses actually saw God firsthand is contradicted by the statement God made to Moses: "But," He said, "you cannot see My face, for man may not see Me and live" (Exod. 33:20).

Because a person cannot be truly tested if God's existence has already been confirmed by some unequivocal revelatory experience, I assume that God most desires from His subjects an expression of belief that relies on faith alone (i.e., belief without evidence). Indeed, it is not unfair, in my opinion, to read the Bible as the almost obsessive testing of human beings by God to distinguish the faithful from those whose commitment to Him is lacking in zeal or persistence (Job's faith faltered, but he never abandoned God, as I discuss in section 7.4).

This all-too-brief justification of SB's goals by way of the biblical God's statements and actions will not be persuasive to those who regard the Bible as an unreliable source at best, pure fantasy at worst.[9] It is *not*, however, a nonbeliever—or, for that matter, a believer—whom I postulate as P in the Belief Game 1. Instead, I assume that P is somebody who takes the Bible (or other monotheistic religious works) seriously. Although these works may describe experiences that are outside P's ken or beyond the secular world, I suppose that P has yet to make up her mind about the existence of an ultimate reality embodied in some SB.[10]

While P entertains the possibility of SB's existence, and in fact would prefer confirmatory to nonconfirmatory evidence in the Belief Game 1 (according to her secondary goal), *evidence is P's major concern* (i.e.,

---

8. McShane (2014) claims otherwise in her revised Belief Game, arguing that God has a dominant strategy of revealing Himself. She goes further, claiming to show that "any game-theoretic model in which a human being and God are players can only succeed at the cost of abandoning the assumption that God is omnibenevolent" (p. 3)—in other words, there is a price to pay for not assuming that God is omnibenevolent. I leave it to the reader to judge whether the biblical evidence demonstrates that God is omnibenevolent.

9. For more evidence on God's goals beyond the cursory biblical citations provided here, see Brams [1980] 2003.

10. "Ultimate reality" is a term used in philosophy of religion and theology to refer to different conceptions or models of God that are used in different religions. For a comprehensive analysis of these models, see Diller and Kasher 2013.

primary goal). Moreover, P realizes that whether or not SB provides it will depend on what SB's rational choice in the game is.

To highlight the quandary that the Belief Game 1 poses for both players, observe that SB has a *dominant strategy* of $\overline{R}$: This strategy is better for SB, whether P selects B (because SB prefers (4,2) to (3,4)) or $\overline{B}$ (because SB prefers (2,3) to (1,1)). Given SB's dominant strategy of $\overline{R}$, P—who does not have a dominant strategy but prefers (2,3) to (4,2) in the second row of the payoff matrix (which is SB's dominant strategy and which P can presume SB will choose in a game of complete information)—will choose $\overline{B}$. That is, $\overline{B}$ is P's best response to SB's choice of $\overline{R}$.

The choice of these strategies by the players leads to the selection of (2,3), which is the unique *Nash equilibrium* and underscored in figure 3.1. It is an outcome from which neither player would depart unilaterally because he or she would do worse doing so: SB would go from 2 to 1 if he switched from $\overline{R}$ to R, and P would go from 3 to 2 if she switched from $\overline{B}$ to B.[11]

Observe that (2,3) is worse for both players than the outcome that would result if SB chose R and P chose B, which would yield (3,4) in Belief Game 1. The latter outcome is *Pareto-superior*, or better for both players, to the Nash equilibrium of (2,3), whereas (1,1) is Pareto-inferior, or worse for both players, not only to (3,4) but also to (2,3). If there is no outcome in a game that is better for *both* players, it is *Pareto-optimal*. Both (3,4) and (4,2) are Pareto-optimal in Belief Game 1, whereas (2,3) and (1,1) are *Pareto-nonoptimal*.

Even though (3,4) is better for both players than (2,3), (3,4) is not a Nash equilibrium, because SB has an incentive, once at (3,4), to depart to (4,2). Neither is (4,2) a Nash equilibrium, because once there P would prefer to move to (2,3). And, of course, both players would prefer to move from (1,1). Thus, if the players have complete information about each other's preferences and choose their strategies independently of each other, game theory predicts the choice of (2,3) that, paradoxically, is worse for both players than (3,4).[12]

---

11. A Nash equilibrium is actually the strategy pair that defines this outcome, not the outcome itself. Henceforth, however, I will usually identify Nash equilibria by the outcomes they yield. This will cause no confusion in 2 × 2 strict ordinal games, because a strategy pair is uniquely defined by each of its four outcomes.

12. This is also the paradox in game theory's most famous game, Prisoners' Dilemma (game 32 in the appendix). Belief Game 1 differs from Prisoners' Dilemma in being asymmetric—the players do not have the same strategic choices, as in a symmetric game—with only one player (SB) having a dominant strategy and the other player (P)

According to the theory of moves (TOM), Belief Game 1 is "cyclic" in a counterclockwise direction, as shown by the arrows in figure 3.1. Because outcomes may be only temporary as players move and countermove around a payoff matrix, I refer to them as *states* in TOM.

Notice that from every state except the Nash equilibrium, (2,3), it is rational for the player with the next move in the counterclockwise cycle to move to a different state to improve his or her payoff. In the case of (2,3), the fact that the SB would prefer not to move in the direction of the arrow to (1,1) creates what I call an *impediment* (indicated by "i" in figure 3.1), but it does not preclude Belief Game 1 from being cyclic.

A 2 × 2 game is *cyclic*—in either a clockwise or counterclockwise direction (it cannot be cyclic in both directions, as shown in Brams 1994, 90–91)—if the player with the next move in the cycle never receives his or her best payoff (4). Thus, from (2,3), SB can move to a less-preferred state, (1,1), but his move is not from his best state (4) but from his next-worst state (2), rendering Belief Game 1 cyclic in a counterclockwise direction.[13]

Because there is one impediment to the players' moves shown in figure 3.1, Belief Game 1 is *moderately cyclic*. A game is *strongly cyclic* if it has no impediments, and it is *weakly cyclic* if it has two impediments (a 2 × 2 game cannot have more than two impediments, one for each player; see Brams 1994, 94). The fact that the Belief Game 1 is moderately cyclic suggests that there will be some "friction" in the counterclockwise movement of the players around the matrix.

Before applying TOM—and, specifically, the concept of moving power—to Belief Game 1, let me clarify SB's choice of $\overline{R}$, which I interpret as "don't establish his existence" (see figure 3.1). From P's perspective, $\overline{R}$ may occur for two distinct reasons: (i) SB does not exist as a player, or (ii) SB exists but does not choose to reveal himself. Not only can P not distinguish between these two reasons for nonrevelation, but even if SB exists, P knows that SB has a dominant strategy of $\overline{R}$ and would, therefore, almost surely choose it in Belief Game 1.

---

having a best response to it. More generally, a two-person normal-form *symmetric game* is one in which the payoffs can be arranged so that the players' ranks along the main diagonal are the same and their ranks along the off-diagonal are mirror images of each other.

13. This game is not cyclic in a clockwise direction, because a move by SB from (4,2) to (3,4), or a move by P from (3,4) to (1,1), requires that the mover move from his or her best state (4) to an inferior one. When this occurs, we say that a game is not cyclic in this direction because of *payoff termination*.

For this reason, I do not assume that P would ever think there is conclusive evidence of *nonexistence*, so I do not give P this option in the Belief Game 1. Instead, P can choose *not to believe* in SB's existence and—though this is not shown in the matrix—not to believe in SB's nonexistence, either, which is to say that P is agnostic. That is, P suspends judgment, which I interpret as a kind of commitment to remain noncommital.[14]

In a sense, a thoughtful agnostic plays Belief Game 1, and perhaps Belief Game 2 that I will analyze in section 3.3, all his or her life, never certain about SB's strategy choice, or even that SB exists. In choosing $\bar{B}$, I interpret P to be saying that she does not believe in either SB's existence or nonexistence *yet*—in other words, she wants to keep her options open.

Should P become a believer or a nonbeliever, then she no longer would be torn by the self-doubt reflected by P's choices in Belief Game 1. The evidence, so to speak, would be in. But I assume that P is neither an avowed theist nor an avowed atheist but a person with a scientific bent, who desires confirmation of either belief or nonbelief. In preferring the former to the latter as a secondary goal in Belief Game 1, P is definitely not an inveterate skeptic.

What SB might desire, on the other hand, is harder to discern. Certainly the God of the Hebrew Bible very much sought, especially from His chosen people, the Israelites, untrammeled faith and demonstrations of it. Although He never revealed Himself in any physical form, except possibly to Moses before he died, He continually demonstrated His powers in other ways, especially by punishing those He considered transgressors.

SB has *moving power* if he can continue moving when P will, eventually, be forced to stop.[15] That is, SB has the endurance or stamina to continue the counterclockwise cycling, whereas P does not and must, at some point, "call it quits."

With moving power, SB can induce P to stop at either (4,2) or (1,1), where P has the next move. P would prefer (4,2), which gives SB his

---

14. Not everyone believes that such openness is desirable, at least in the case of God. For example, Hanson (1971, 303–311) thinks that the proper position of the agnostic on the question of God's existence should be one of reasonable doubt. For Hanson, moreover, the evidence is tipped decisively against God's existence. His view is echoed in a host of more recent books, including Dawkins 2006, Dennett 2006, and Hitchens 2007.

15. Moving power is more formally developed and analyzed in Brams 1994, chap. 5; here I illustrate its application to Belief Game 1 and other games later. Kilgour and Zagare (1987) introduced a related notion, "holding power," which they applied to the 1948 Berlin crisis between the Western powers and the Soviet Union.

best payoff: P's belief without evidence satisfies both of SB's goals. But P obtains only her next-worst payoff in this state; it satisfies only her secondary goal of believing in SB's existence, but not her primary goal of having evidence to support her belief.

Endowing SB with moving power raises a feasibility question about the ability of P to move. Whenever P wavers between belief and non-belief, I assume that SB can switch back and forth between revelation and nonrevelation. But once SB has established his existence, can it be denied? I suggest that this is possible, but only if one views Belief Game 1 as a game played over an extended period of time.

To illustrate this point, consider the situation recounted in chapter 19 of Exodus. After God "called Moses to the top of the mountain" (Exodus 19:20) to give him the Ten Commandments, there was "thunder and lightning, and a dense cloud ... and a very loud blast of the horn" (Exodus 19:16). This display provided strong evidence of God's existence to the Israelites, but for readers of the Bible today, it is perhaps not so compelling.

However, the Israelites became wary and restive after Moses's absence on Mount Sinai for forty days and nights (see section 9.5). With the complicity of Aaron, Moses's brother, they revolted and built themselves a golden calf. God's earlier displays of might and prowess had lost their immediacy and therefore their force.

This insurrection enraged Moses and God. Moses destroyed the Ten Commandments and, with God's assistance, provoked the slaughter by the Levites (a tribe of Israelites) of 3,000 other Israelites for their idolatry. Moses was a Levite, and he rallied his fellow tribesmen to his side to carry out the slaughter, presumably to demonstrate to God that the idolatry of His chosen people would not go unpunished.

Jumping ahead to the present, the basis of belief for agnostics would seem even more fragile. Those who seek a revelatory experience from reading the Bible may find it, but many do not. For them, God remains hidden or beyond belief unless they can apprehend Him in other ways.

This is how the problem of revelation arises. Without a personal revelatory experience, or the reinforcement of one's belief in God that may come from reading a religious work or going to religious services, belief in God's existence may be difficult to sustain with unswerving commitment.

Revelation, also, may be a matter of degree. If God appears with sound and fury, as He did at Mount Sinai, He may likewise disappear like the morning fog as memories of Him slowly fade. Thereby seeds of doubt are planted. But a renewal of faith may also occur if a person experiences some sort of spiritual awakening.

A wavering between belief and nonbelief created by SB's moving between revelation and nonrevelation shows that P's belief in SB *may have a rational basis for being unstable*. Sometimes the evidence manifests itself, sometimes not, in Belief Game 1. What is significant in this cyclic game is that SB's exercise of moving power is consistent with SB's sporadic appearance and disappearance—and with P's intermittently responding to revelation by belief, to nonrevelation by nonbelief.

In the Bible, God wants to remain inscrutable, as the following colloquy, which I alluded to in the preface, suggests:

> Moses said to God, "When I come to the Israelites and say to them 'The God of your fathers has sent me to you,' and they ask me, 'What is His name?' what shall I say to them?" And God said to Moses, "Ehyeh-Asher-Ehyeh" ["I Am That I Am"]. He continued, "Thus shall you say to the Israelites, 'Ehyey [I Am] sent me to you.'" (Exod. 3:13–14)

As enigmatic as this reply was, however, God was also quick to trumpet His deeds and demonstrate His powers, as when He said in Moses's confrontation with Pharaoh (see section 9.3) that He would "multiply My signs and marvels in the land of Egypt" (Exod. 7:4).

By inflicting ten plagues upon the Egyptians, God induced Pharaoh to release the Israelites from their bondage. When the Egyptians pursued the Israelites to the Red Sea, which had parted for them, God locked the chariot wheels of the Egyptians as they crossed the Sea, causing them all to drown when the Sea came together again, swallowing them up.

Relying on faith alone, when reason does not sustain belief, produces an obvious tension in P. Over a lifetime, P may oscillate between belief and nonbelief as seeming evidence appears and disappears. For example, the indescribable tragedy of the Holocaust destroyed the faith of many believers, especially Jews, in a benevolent God, and for some it will never be restored.

But for others it has been restored. Furthermore, many former nonbelievers have conversion experiences—sometimes induced by mystical episodes—and, as a result, pledge their lives to some deity. For still others, there is a more gradual drift either toward or away from religion and a belief in an SB.

More broadly, there are periods of religious revival and decline, which extend over generations and even centuries, that may reflect a collective consciousness about the presence or absence of an SB—or maybe both. As Kolakowski (1982, 140) remarked, "The world manifests God and conceals Him at the same time."

It is, of course, impossible to say whether an SB, operating behind the scenes, is ingeniously plotting his moves in response to the moves, in one direction or another, of individuals or societies. For many today, this is an Age of Reason, which has had different names in the past (for example, Age of Enlightenment), in which people seek out rational explanations for perplexing phenomena. But there is also evidence of a religious reawakening (for example, as occurred during the Crusades), which often follows periods of secular dominance. This ebb and flow is inherent in the instability of moves in Belief Game 1, even if an SB, possessed of moving power, has his way on occasion and is able temporarily to implement outcome (4,2).

Perhaps the principal difficulty for SB in making this outcome stick is that peoples' memories erode after a prolonged period of nonrevelation. Consequently, the foundations that support belief may crumble. Nonbelief sets up the need for some new revelatory experiences, sometimes embodied in a latter-day messiah, followed by a rise and then another collapse of faith.

If P, perhaps implausibly, is treated as the player who possesses moving power, then she can induce (3,4) by forcing SB to choose between (3,4) and (2,3), given that SB must stop at one of those states when it has the next move. Because SB prefers (3,4), he will stop at this state, which is P's best state and the outcome she can induce with moving power.

We say that moving power is *effective* if the player who possesses it can induce a better outcome than would be induced if the other player possessed it. In Belief Game 1, P and SB can induce their best states— (3,4) and (4,2), respectively—when each exercises moving power, so they obviously would prefer to have it than for the other player to have it.

If the idea of P's "forcing" SB to reveal himself at (3,4)—and, on this basis, for P to believe—sounds absurd, it is useful to recall that the biblical God exerted Himself mightily to demonstrate His awesome powers to new generations, which can be interpreted as a form of revelation. On occasion, however, God would leave the stage in order to test a new generation's faith. But after a period of nonrevelation, He would usually be forced to return in order to try to foster belief again.

The effects of moving power, whether possessed by SB or P, seem best interpreted in Belief Game 1 as occurring over extended periods of time. Memories fade, inducing SB to move from nonrevelation to revelation when the next generation does not understand or appreciate SB's earlier presence. Even when SB moves in the opposite direction, going from revelation to nonrevelation, his actions may not appear inconsistent if P, effectively, is a different player. Thereby the earlier concern I raised about infeasible moves is dissipated in an extended game in which the identity of P, and memories she had, change. Over time, therefore, a different P may vacillate between belief and nonbelief.

3.3   Belief Game 2

My treatment of Belief Game 2 will be brief, because the goals and preferences of SB do not change from those of Belief Game 1. Neither does the (i) primary goal of P, which I repeat below, but her (ii) secondary goal does change:

P: (i) wants belief (or nonbelief) in SB's existence confirmed by evidence (or lack thereof); (ii) prefers *not* to believe in SB's existence.

Recall that P's primary goal means that her two best outcomes (3 and 4) occur along the main diagonal—when revelation by SB supports belief by P, and nonrevelation supports nonbelief—and her two worst outcomes (1 and 2) occur along the off-diagonal, when there is not supporting positive or negative evidence for either belief or nonbelief by P. However, P's new secondary goal means that her preferred outcomes in each of the preceding pairs (4 in the first and 2 in the second) are now associated with $\bar{B}$ instead of B (see figure 3.2). Belief Game 2 is game 49 in the appendix.

What effect does P's preference for nonbelief have when there is nonrevelation—as opposed to belief when there is revelation—on play of Belief Game 2 in figure 3.2? First, observe that SB still has a dominant strategy of $\bar{R}$, which is better for him regardless of the strategy that P chooses. P does not have a dominant strategy—her better choice depends on what SB does—but anticipating SB's choice of $\bar{R}$, she will choose $\bar{B}$, which yields (2,4), the unique Nash equilibrium.

Note that (2,4) is the best outcome for P, compared with (2,3) in Belief Game 1. Moreover, $\overline{RB}$ is no longer Pareto-inferior to RB—both (2,4) and (3,3) are Pareto-optimal—which was not true in Belief Game 1. This

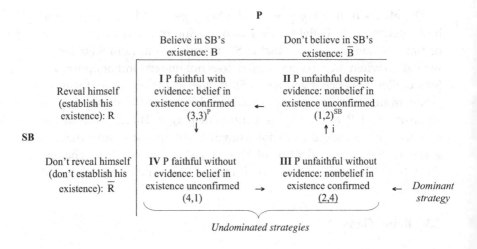

P

	Believe in SB's existence: B	Don't believe in SB's existence: $\bar{\text{B}}$
Reveal himself (establish his existence): R	**I** P faithful with evidence: belief in existence confirmed $(3,3)^P$ ↓	**II** P unfaithful despite evidence: nonbelief in existence unconfirmed $(1,2)^{SB}$ ↑ i
Don't reveal himself (don't establish his existence): $\bar{\text{R}}$	**IV** P faithful without evidence: belief in existence unconfirmed $(4,1)$ →	**III** P unfaithful without evidence: nonbelief in existence confirmed $\underline{(2,4)}$

SB (label at left)

← (arrow between I and II)

← *Dominant strategy*

Undominated strategies

Key: (x,y) = (payoff to SB, payoff to P)
   4 = best; 3 = next-best; 2 = next-worst; 1 = worst
   i = impediment
   Nash equilibrium underscored
   Arrows indicate progression of states in moderately cyclical game
   Superscripts P and SB indicate outcomes each player can induce with moving power

Figure 3.2
Outcome and payoff matrix of Belief Game 2 (game 49)

enhances (2,4)'s status as not just a stable outcome but also one that both players do not consider inferior to another outcome, as was (2,3) in Belief Game 1.

What one might view as the cooperative outcome, (3,3), in Belief Game 2—at which both players obtain their next-best choices—is still not stable (that is, a Nash equilibrium), as was true of (3,4) in Belief Game 1: Once there, SB has an incentive to defect to (4,1), which is P's worst outcome. Furthermore, this game, like Belief Game 1, is cyclic, because neither player ever departs from his or her best outcome when players move and countermove in a counterclockwise direction. In fact, also like Belief Game 1, Belief Game 2, is moderately cyclic, because it has one impediment—when SB moves from (2,4) to (1,2). Thus, it has the same kind of instability, with some friction, as Belief Game 1 does.

The effects of moving power, however, are curious in Belief Game 2. When P possesses it, she can force SB to stop at either (3,3) or (2,4), where he has the next move, so he would choose (3,3), the cooperative outcome. When SB has moving power, he can force P to stop at either (4,1) or (1,2), where she has the next move, so she would choose (1,2). Because (1,2) is Pareto-inferior to (3,3), moving power is *irrelevant* in

Belief Game 2—both players would prefer to stop at (3,3). Consequently, even if SB had moving power, he would not exercise it but instead would stop the cycling at (3,3).

This contrasts with SB's exercise of moving power in Belief Game 1, wherein he can induce his best outcome, (4,2). Because moving power is effective in Belief Game 1, however, P also can do better if she possesses it, inducing her own best outcome of (3,4) (see section 3.2). But in Belief Game 2, SB would be foolish to induce (1,2), preferring to stop at (3,3).

In summary, Belief Games 1 and 2 share several properties: SB has a dominant strategy of $\overline{R}$, which leads to the Nash equilibrium of $\overline{RB}$. But unlike Belief Game 1, this nonrevelation-nonbelief outcome is not Pareto-inferior to RB in Belief Game 2. In fact, $\overline{RB}$, yielding (2,4), is P's best outcome in Belief Game 2: She is happiest in not believing when SB does not reveal himself, justifying her nonbelief.

But (2,4) is SB's next-worst state, so he will not be pleased at it for long. On the one hand, his moving power, if he possesses it, is no help, because if he forces P to stop when she has the next move, it would be at (1,2), where SB does even worse. On the other hand, P can induce (3,3) with moving power, which presumably SB would accept if the game cycles and he wants it to stop.

But then SB can do even better by moving to (4,1), which is disastrous for P. So P would move on to (2,4), the Nash equilibrium and P's best state. Now the players are back to "square one," which I previously showed is not likely to stay stable, despite its being a Nash equilibrium.

I interpret outcomes in Belief Games 1 and 2 not as final but instead as states that the players pass through, sometimes staying and sometimes moving on. Players move on either because it is not rational for them to stay or because one player, if he or she possesses moving power, can force stoppage of the player without moving power, although invoking this power is not always optimal, as I showed for SB in Belief Game 2.

Fundamentally, both Belief Games are games for the ages, especially with the identity of P constantly in flux. Their fluidity, in my opinion, is their most striking feature.

3.4 Conclusions

In this chapter and in chapter 2, I have analyzed vexing theological questions that inhere in two different kinds of choice situations:

• *decisions* in chapter 2, in which states of nature arise according to some chance mechanism; P alone makes choices without ever being certain in which state she is in;
• *games* in chapter 3, in which both P and SB are active players with preferences; both make choices in order to try to obtain preferred outcomes.

In the case of decisions, however, I did allow for the possibility that SB might be able to induce certain states of nature in order to advance his goals, making the decision for P more game-like and not just a matter of "playing the odds."

For plausible assumptions about SB's and P's preferences in Belief Games 1 and 2, I showed that SB has a dominant strategy of nonrevelation, which induces P not to believe in SB's existence. This strategy pair gives the unique Nash equilibrium of (2,3) in Belief Game 1 and (2,4) in Belief Game 2.

Unfortunately for the players in Belief Game 1, this outcome is Pareto-inferior to (3,4), wherein SB reveals himself and P believes, which P can induce with moving power. But it seems more likely that SB, if he is immortal, is the player who possesses moving power, enabling him to induce (4,2), wherein he does not reveal himself yet P believes.

On occasion, however, it seems not beyond the pale that P can hold out longer and, with moving power, induce her most-preferred state, (3,4). Especially when P assumes new identities over time and cannot recall the past, she is more likely to continue to cycle, unresponsive to the lessons SB intended her to learn earlier. The counterclockwise cycling around Belief Game 1 suggests why we go through periods of religious revival and decline over the ages.

The situation changes in Belief Game 2, wherein P prefers nonrevelation-nonbelief ($\overline{RB}$), giving (2,4), to revelation-belief (RB), giving (3,3). With moving power, P can induce (3,3), which, paradoxically, is better for both players than (1,2) that SB can induce with moving power.

Presumably, therefore, SB would accede to (3,3) and stop there in the move-countermove process. If SB instead chooses his dominant strategy of $\overline{R}$ and refuses to budge from it, he could end up at a worse state—the Nash equilibrium, (2,4)—and regret his choice.

Neither Pascal's wager nor the Search and Concern Decisions offers insight into why cycling occurs. On the other hand, a game-theoretic perspective, supplemented by TOM—and, in particular, the notion of moving power—does. It also clarifies why the answer that Pascal's wager gives for believing in God may not always be persuasive when God has preferences and is capable of making His own choices.

In chapter 4 I will define a farsighted concept of equilibrium, "non-myopic equilibrium," whereby players think ahead about the possible moves of another player, as well as their own best moves, before making choices. This will allow us to define a new concept of stability, different from that of a Nash equilibrium, as well as explore other notions of power besides moving power.

4   Nonmyopic Equilibria in the Belief Games

4.1   Introduction

In chapter 3 I introduced the concept of moving power, which benefits the player who possesses it in Belief Game 1 by enabling him or her to induce a best outcome. As shown in figure 3.1, SB can induce (4,2), whereas P can induce (3,4), rendering moving power effective in this game.

By comparison, the standard solution concept in game theory, Nash equilibrium, gives (2,3) as the outcome in Belief Game 1, in which SB does not reveal himself and P does not believe. This outcome is Pareto-inferior to (3,4) but not (4,2), at which P does even worse than (2,3) if SB has moving power.

Unfortunately for SB, this kind of power does not help him in Belief Game 2 in figure 3.2. Recall from chapter 3 that if SB has moving power and tries to induce his best outcome of (4,1) by forcing P to stop at her preferred outcome when she has the next move, SB succeeds only in inducing his worst outcome, (1,2). It would be better for SB to stop, on his own volition, at (3,3), when he has the next move.

SB's "problem" in Belief Game 2 is that his best outcome is P's worst, (4,1), so P has no reason to stop at it as the players cycle counterclockwise in this game (see figure 3.2). P would do better if play stopped at any other outcome, including (3,3), which with moving power she can induce, though even better for her would be the Nash equilibrium, (2,4).

In section 3.3, outcome (3,3) is what I called "cooperative." It is not only Pareto-optimal but also next best for both players. However, it is not a Nash equilibrium, so game theory does not single it out as stable.

SB's moving power also does not single (3,3) out—not to mention (4,1), SB's best outcome—so it would be impossible to identify SB as superior by the outcome that he can implement with moving power. To be sure, there are other kinds of power, which we will analyze later, as possible ways by which SB can signal, and even demonstrate, his superiority.

Instead of trying to identify SB by the outcomes he can implement, I assume in this chapter that SB and P make rational calculations, according to TOM, about what outcomes they will choose from any initial state in a game. More specifically, the players think ahead about their moves and countermoves in light of the moves and countermoves they anticipate the other player will make, selecting those that they prefer (in a manner to be spelled out in section 4.2).

To determine where they will end up, we need a concept of equilibrium different from Nash's concept, which assumes that players think only one step ahead: Can I do better by moving or staying? If the answer is "staying," the outcome is in equilibrium.[1] By contrast, the concept of a "nonmyopic equilibrium," which I define and illustrate in section 4.2, assumes that the players look several steps ahead before making a decision.

Nash equilibria are most applicable to the thinking of nonhuman animals, who typically respond myopically to a dangerous situation either by fighting or fleeing (sometimes, of course, they freeze in fear, as do human beings). But human beings, and I presume superior beings, can and do think more than one step ahead about the consequences of their choices. To incorporate this more farsighted thinking into the play of games in which I do not assume endless cycling or the exercise of moving power, we need to specify new rules of play, which I do in section 4.2.

These rules assume the ability of players to make calculations based on a central tenet of game theory, backward induction. Backward induction enables one to calculate nonmyopic equilibria and show that every 2 × 2 game has at least one such equilibrium. Because Belief Games 1 and 2 each have two, there is uncertainty about which, if either, will be chosen. Moreover, although one equilibrium is common to both games—involving revelation and belief—the second equilibrium in each game differs qualitatively from its counterpart.

---

1. A notion of equilibrium, proposed by Stackelberg ([1934] 1952), supposes that players look two steps ahead by assuming that one player (the leader) acts before the other player (the follower) does. More precisely, a *Stackelberg equilibrium* is an outcome at which one player (the leader), anticipating the best response of the other player (the follower) to its (the leader's) choice of a strategy, cannot obtain a better outcome for itself by choosing a different strategy. Related notions are analyzed in Zagare 1984.

In Belief Game 2, neither of the equilibria is stable in the following sense: Once one is reached, the players have an incentive to move to the other (this does not occur in Belief Game 1), creating a boomerang effect. This back-and-forth movement renders the players' choices constantly in flux, which provides a rational basis for the instability due to cycling that I discussed in chapter 3.

One consequence of this instability is that it is hard to determine whether there is any rhyme or reason to SB's choices—in particular, whether he is trying to implement an outcome that would satisfy his goals and, possibly, make his superiority evident. In other words, the boomerang effect, and its seemingly random choices, leave ambiguous what makes SB superior.

4.2 Nonmyopic Equilibria

In this chapter, the starting point is a payoff matrix, or *game configuration*, in which the order of play is not specified. By contrast, in chapter 3 the order was specified in cyclic games: Moves could be either clockwise or counterclockwise.

Unlike chapter 3, I do not restrict the analysis to cyclic games. In every 2 × 2 game, players may move in either direction from outcomes by looking ahead and using backward induction to determine the rationality of both their moves and those of another player (or, more generally, other players).

Because standard game theory assumes that players choose strategies simultaneously in games in normal or strategic form (i.e., given by a payoff matrix),[2] it does not raise questions about the rationality of moving or departing from outcomes—at least beyond an immediate departure, à la Nash. In fact, however, most real-life games do not start with simultaneous strategy choices but instead commence at an outcome, which is the status quo that initiates play of a game;

---

2. Strategies may allow for sequential choices, but game theory models, in general, do not make endogenous who moves first, as TOM does, but instead specify a fixed order of play (i.e., players make either simultaneous or sequential choices). There are exceptions in the literature, however, including Hamilton and Slutsky 1988, 1990; Rosenthal 1991; Amir 1995; and van Damme and Hurkens 1996. Typically, these models allow a player in the preplay phase of a game to choose when he or she will move in the play of the game. Yet the choice of when to move applies only to a player's initial strategy choice, whereas the nonmyopic calculations to be described next assume that players, starting at states, make moves and countermoves that depend on thinking several steps ahead. I do not attempt a full technical exposition here, which can be found, with examples, in Brams 1994.

moreover, it is not a random choice made by nature, as is often assumed in decision theory and some game theory models, but is instead the product of previous choices. The question then becomes whether a player, by departing from the status quo, can do better not only in an immediate or myopic sense but also in an extended or non-myopic sense.

In the case of 2 × 2 games, TOM postulates four *rules of play*, which describe the possible choices of the players at different stages:

1. Play starts at an outcome, called the *initial state*, which is at the intersection of the row and column of a 2 × 2 payoff matrix.

2. Either player can unilaterally switch his or her strategy, and thereby change the initial state into a new state, in the same row or column as the initial state.[3] The player who switches, who may be either row (R) or column (C), is called player 1 (P1).

3. Player 2 (P2) can respond by unilaterally switching his or her strategy, thereby moving the game to a new state.

4. The alternating responses continue until the player (P1 or P2) whose turn it is to move next chooses not to switch his or her strategy. When this happens, the game terminates in a *final state*, which is the *outcome* of the game.

Note that the sequence of moves and countermoves is strictly alternating: First, say, R moves, then C moves, and so on, until one player stops, at which point the state reached is final and, therefore, the outcome of the game.[4]

The use of the word "state" is meant to convey the temporary nature of an outcome, before players decide to stop switching strategies. I assume that no payoffs accrue to players from being in a state unless

---

3. I do not use "strategy" in the usual sense to mean a complete plan of responses by the players to all possible contingencies allowed by rules 2–4, because this would make the normal form, as represented by a payoff matrix, unduly complicated to analyze. Rather, *strategies* refer to the choices made by players that define an initial state, and *moves and countermoves* refer to the players' subsequent strategy switches from an initial state to a final state in an extensive-form game represented by a game tree, as allowed by rules 2–4. For another approach to combining the normal and extensive forms, see Mailath, Samuelson, and Swinkels 1993, 1994.

4. An emendation in the rules of TOM that allows for backtracking would be appropriate in games of incomplete information, wherein players may make mistakes that they later wish to rectify. Implications of allowing backtracking on nonmyopic equilibria, which are discussed in Brams 1994, are analyzed in Willson 1998 and applied in Zeager, Ericson, and Williams 2013.

it is the final state and, therefore, becomes the outcome (which could be the initial state if the players choose not to move from it).

Rule 1 differs radically from the corresponding rule of play in standard game theory, in which players simultaneously choose strategies in a matrix game, which determines its outcome. Instead of starting with strategy choices, I assume that players are already in some state at the start of play (the status quo) and receive payoffs from this state only if they choose to stay. Based on these payoffs, they decide, individually, whether or not to change this state in order to try to do better.[5]

To be sure, some decisions are made collectively by players, in which case it would be reasonable to say that they choose strategies from scratch, either simultaneously or by coordinating their choices. But if, say, two countries are coordinating their choices, as when they agree to sign a treaty, the most important strategic question is what individualistic calculations led them to this point. The formality of jointly signing the treaty is the culmination of their negotiations, which does not reveal the move-countermove process that preceded it. This is precisely what TOM is designed to uncover.

In summary, play of a game starts in a state, at which players accrue payoffs only if they remain in that state so that it becomes the outcome of the game. If they do not remain there, they still know what payoffs they would have accrued had they stayed; hence, they can make a rational calculation of the advantages of staying versus moving. They move precisely because they calculate that they can do better by switching states, anticipating a better outcome when the move-countermove process finally comes to rest.

Rules 1–4 say nothing about what causes a game to end, but only when: Termination occurs when a "player whose turn it is to move next chooses not to switch its strategy" (rule 4). But when is it rational not to continue moving, or not to move in the first place from the initial state?

To answer this question, I posit a rule of *rational termination* (first proposed in Brams [1983] 2007, 106–107), which has been called

---

5. Alternatively, players may be thought of as choosing strategies initially, after which they perform a thought experiment of where moves will carry them once a state is selected. The concept of an "anticipation game," developed in section 4.4, advances this idea, which might be considered dynamic thinking about the static play of a matrix game. Generally, however, I assume that "moves" describe actions, not just thoughts, though I certainly allow for the possibility of a thought interpretation or thought experiment (*Gedankenexperiment* in German).

"inertia" by Kilgour and Zagare (1987, 94). It prohibits a player from moving from an initial state unless doing so leads to a better (not just the same) final state according to this rule:

5. A player will not move from an initial state if this move (i) leads to a less preferred final state (i.e., outcome); or (ii) returns play to the initial state (i.e., makes the initial state the outcome).

I discuss and illustrate shortly how rational players, starting from some initial state, determine by backward induction what the outcome will be.

Condition (i) of rule 5, which precludes moves that result in an inferior state, needs no defense. But condition (ii), which precludes moves that will cause players to cycle back to the initial state, is worth some elaboration. It says that if it is rational for play of the game to cycle back to the initial state after Pl moves, Pl will not move in the first place. After all, what is the point of initiating the move-counter-move process if play simply returns to "square one," given that the players receive no payoffs along the way (i.e., before an outcome is reached)?

Not only is there no gain from cycling but, in fact, there may be a loss because of so-called transaction costs—including the psychic energy spent—that players suffer by virtue of making moves that, ultimately, do not change the situation.[6] Therefore, it seems sensible to assume that P1 will not trigger a move-countermove process if it only returns the players to the initial state, making it the outcome.

I call rule 5 a *rationality rule*, because it provides the basis for players to determine whether they can do better by moving from a state or remaining in it. Still another rationality rule is needed to ensure that both players take into account each other's calculations before deciding to move from the initial state. I call this rule the *two-sidedness rule*:

6. Given that players have complete information about each other's preferences and act according to the rules of TOM, each takes into account the consequences of the other player's rational choices, as well as his or her own, in deciding whether to move from the initial

---

6. The rules of play that allow for cycling, which I discussed in chapter 3, violate this prohibition on cycling. Whereas the cycling rules enable one to analyze the effects of moving power, which is assumed to be possessed by one player, the present rules treat SB and P symmetrically. However, I will consider other forms of power later to determine whether, if SB or P possesses such power, he or she can either reinforce or upset certain nonmyopic equilibria.

state or subsequently, based on backward induction (to be defined and illustrated in section 4.3). If it is rational for one player to move and the other player not to move from the initial state, then the player who moves takes *precedence*: His or her move overrides the player who stays, so the outcome is that induced by the player who moves.

Because players have complete information, they can look ahead and anticipate the consequences of their moves. I next show how, using backward induction, SB and P can do this in Belief Games 1 and 2.

4.3 Nonmyopic Equilibria in Belief Games 1 and 2

In Belief Game 1, suppose that play commences at the Nash equilibrium, (2,3), from which SB and P have no immediate incentive to move (see figure 4.1). Because SB receives only his next-worst payoff, however, he, especially, would have good reason to calculate that even though moving would bring him to his worst state of (1,1), this is also P's worst

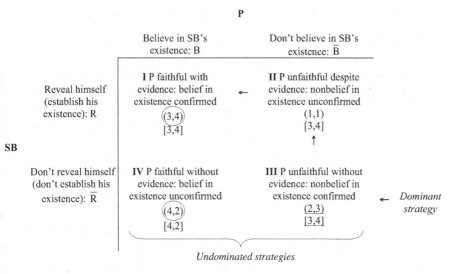

Key: (x,y) = (payoff to SB, payoff to P)
   4 = best; 3 = next-best; 2 = next-worst; 1 = worst
   Anticipation game given in brackets
   Nash equilibria underscored
   Nonmyopic equilibria (NMEs) circled
   Arrows show movement to NME of (3,4) from (2,3)

Figure 4.1
Nonmyopic equilibria in Belief Game 1 (game 48)

state. This would give P an incentive to countermove if SB moves to (1,1), thereby bringing the players to (3,4), which is Pareto-superior to (2,3).

But then would SB counter-countermove to (4,2), and P counter-counter-countermove back to the initial state of (2,3)? What we need to determine are the rational consequences of moving from an initial state—in particular, what moves and countermoves would be triggered if SB moves from (2,3) to (1,1) in Belief Game 1.

I assume that the players base their calculations on *backward induction*, which I describe next. Starting from (2,3) and cycling counterclockwise back to this state, I next show *where* the row player, R (i.e., SB), and the column player C (i.e., P), will terminate play.

If R moves first, the counterclockwise progression is from (2,3) back to (2,3). The player (R or C) who makes the next move, shown below each state, alternates:[7]

	State 1	State 2	State 3	State 4	State 1
	R	C	R	C	
R starts:	(2,3) →	(1,1) →	(3,4) →|	(4,2) →	(2,3)
Survivor:	(3,4)	(3,4)	(3,4)	(2,3)	

The *survivor* is determined by working backward, after a putative cycle has been completed, which is calculated in the following manner. Assume that the players' alternating moves have taken them counterclockwise from (2,3) to (1,1) to (3,4) to (4,2), at which point C must decide whether to stop at (4,2) or complete the cycle and return to (2,3). Clearly, C prefers (2,3) to (4,2), so (2,3) is listed as the survivor below (4,2): Because C *would* move the process back to (2,3) should it reach (4,2), the players know that if the move-countermove process reaches this state, the outcome will be (2,3).

Knowing this, would R at the prior state, (3,4), move to (4,2)? Because R prefers (3,4) to the survivor at (4,2)—namely, (2,3)—the answer is no. Hence, (3,4) becomes the survivor when R must choose between stopping at (3,4) and moving to (4,2), which, as I just showed, would become (2,3) once (4,2) is reached.

---

7. Effectively, this is a *game tree*, or *game in extensive form*, showing a sequence of alternating choices of the players, except that instead of branching from top to bottom, as is the usual representation, the choices of the players go sideways, from left to right. More conventional game trees that illustrate TOM calculations are given in Taylor and Pacelli 2008.

At the prior state, (1,1), C would prefer moving to (3,4) rather than stopping at (1,1), so (3,4) again is the survivor if the process reaches (1,1). Similarly, at the initial state, (2,3), because R prefers the previous survivor, (3,4), to (2,3), (3,4) is the survivor at this state as well. I underscore (3,4) at the point that it becomes the survivor state.

The fact that (3,4) is also the survivor at the initial state, (2,3), means that it is rational for R initially to move to (1,1), and C subsequently to move to (3,4), where the process will stop, making (3,4) the rational choice if R has the opportunity to move first from initial state (2,3). (I will shortly give the backward-induction calculations from (2,3) if C moves first.) That is, after working *backward* from C's choice ofcompleting or not completing the cycle at (4,2), the players can reverse the process and, looking *forward*, determine that it is rational for R to move from (2,3) to (1,1), and C subsequently to move from (1,1) to (3,4), at which point R will stop the move-countermove process at (3,4).

Observe that R does better at (3,4) than at (2,3), where it could have terminated play at the outset, and C does better at (3,4) than at (1,1), where it could have terminated play, given that R is the first to move. In addition, I indicate that it is not rational for R to move on from (3,4) by the vertical line blocking the arrow emanating from (3,4), which I refer to as *blockage*: A player will always stop at a blocked state, wherever it is in the progression. *Stoppage* occurs when blockage occurs for the *first* time from some initial state (because it occurs only once in the present example, the blockage point is also the stoppage point).

What happens if C (i.e., P) is able to move first from (2,3)? Would it do so, and if so to where would play stop? I give the backward-induction calculations below, starting at (3,2), but now moves and countermoves proceed in a clockwise direction when C initiates the process:

	State 1		State 2		State 3		State 4		State 1
	C		R		C		R		
C starts:	(2,3)	→|	(4,2)	→|	(3,4)	→	(1,1)	→	(2,3)
Survivor:	(2,3)		(4,2)		(3,4)		(2,3)		

As when R has the first move, (2,3) is the first survivor, working backward from the last possible move by R at (1,1), but then it is displaced by (3,4), which C prefers. Next, because R at (4,2) prefers this state to (3,4), (4,2) becomes the new survivor. Finally, (4,2) is displaced by (2,3) at the initial state, which becomes the last survivor and is underscored.

Although the first blockage from the left and, therefore, stoppage occurs at (2,3), blockage would occur subsequently at (4,2) if, for any reason, stoppage does not terminate moves at (2,3). In other words, if C moved initially to (4,2), R would then be blocked. Hence, blockage occurs at two states when C starts the move-countermove process, whereas it occurs only once when R does.

The fact that the rational choice depends on which player has the first move—(3,4) is rational if R starts, (2,3) if C starts—leads to a conflict over which outcome will be selected when the process starts at (2,3). However, because it is not rational for C to move from the initial state because it would do worse than if R moved first, R's move takes precedence, according to rule 6, and overrides C's decision to stay. Consequently, when the initial state is (2,3), the outcome will be (3,4), with movement first through the mutually worst state of (1,1) to reach (3,4).

The outcome into which a state goes is called the *nonmyopic equilibrium* (NME) from that state. NMEs may be viewed as the consequence of both players' looking ahead and making rational calculations of where the move-countermove process will transport them—if they move at all, based on the rules of TOM—from each of the four possible initial states. The NMEs in Belief Game 1 are circled in figure 4.1.

To summarize, TOM leads to a unique prediction of (3,4) if play starts in state (2,3), even though (2,3) is the Nash equilibrium according to standard game theory (see section 3.2). If play starts from each possible state in Belief Game 1, analogous backward calculations give the NMEs from these states.

In figure 4.1, the NMEs are shown in brackets below the four possible states. These payoffs define the *anticipation game* of Belief Game 1, because the players can anticipate where they will end up from each possible state in Belief Game 1—either because they stay, or move and countermove, from these states.

The anticipation game shows that wherever the players start, play will bring them to (3,4)—unless they start at (4,2), where they will stay. These are the two NMEs in Belief Game 1, which are circled in figure 4.1 and shown in brackets from each initial state.

If the players do not start in any state but, instead, choose strategies (R or $\overline{R}$ for SB, B or $\overline{B}$ for P) that place them in one of the four states, then the anticipation game shows that SB has a weakly dominant strategy of $\overline{R}$:[8]

• if P chooses B, $\overline{R}$ gives him a payoff of 4, compared with 3 if he chooses R;
• if P chooses, $\overline{B}$, $\overline{R}$ gives him a payoff of 3, which ties with 3 if he chooses R.

P also has a weakly dominant strategy of $\overline{B}$:

• if SB chooses R, $\overline{B}$ gives her a payoff of 4, which ties with 4 if she chooses B;
• if SB chooses $\overline{R}$, $\overline{B}$ gives her a payoff of 4, compared with 2 if she chooses B.

These weakly dominant strategies lead to the Pareto-inferior state of (2,3), from which the moves and countermoves of TOM would bring the players to the Pareto-superior outcome of (3,4).

In applying TOM, we assume that players do not, *de novo*, independently choose strategies in the anticipation game but instead are already situated in some state (that is, the status quo) in the original game, from which they then consider possible moves. If that state is (4,2), wherein SB does not reveal himself and P nevertheless believes, then P receives only her next-worst payoff. Would she move to (2,3), where she would benefit from her next-best state, and benefit even more if the game passes through (1,1) and goes on to (3,4)?

Surprisingly, the answer is no according to TOM. To understand why, consider the backward-induction process, starting from (4,2), when C (i.e., P) moves first:

	State 1	State 2	State 3	State 4	State 1
	C	R	C	R	
C starts:	(4,2) →|c	(2,3) →	(1,1) →	(3,4) →	(4,2)
Survivor:	(4,2)	(4,2)	(4,2)	(4,2)	

---

8. A dominant strategy, as used previously, is a *strongly dominant strategy* if it is always better, whatever strategy the other player chooses. A *weakly dominant strategy* is a strategy that is at least as good, and only sometimes better; in the case of $\overline{R}$, it is better if P chooses B but ties with R if P chooses $\overline{B}$, as shown in the text. Similarly, P has a weakly dominant strategy of $\overline{B}$.

Notice that (4,2) is the survivor at every state, so if C moves first from (4,2), the process would cycle back to (4,2).

When this happens, then by condition (ii) of rule 5 a player will not initiate a move from the initial state. Hence, C would stay at (4,2), which I indicate by "c" (for cycling) next to C's blockage at this state. (This might be interpreted as a special kind of blockage.) In Belief Game 1, therefore, if play starts out at (4,2), P—as well as SB, who receives his best payoff at this state—will not move from it, despite the fact that P receives only her next-worst payoff at this state.

Notice that both NMEs in Belief Game 1 are associated with B. One would expect, therefore, that P will end up at one of her belief states: (3,4) if the initial state is either (3,4) or one of the two nonbelief states, (1,1) or (2,3); (4,2) otherwise, which in fact SB prefers to (3,4) since he does not need to reveal himself. The propensity of P to believe at both NMEs in Belief Game 1 stems, in part, from her secondary goal of preferring belief to nonbelief.[9]

I now turn to Belief Game 2, wherein a more skeptical P prefers nonbelief when there is no revelation, (2,4), to belief when there is revelation, (3,3), which makes P's belief in SB more difficult to sustain (see figure 4.2). In fact, without revelation ($\overline{R}$), the NME at $\overline{R}B$ in Belief Game 1, (4,2), is replaced by the NME, (2,4), so it is SB rather than P that does poorly starting from this state in Belief Game 2.

In both Belief Games, SB has a dominant strategy of $\overline{R}$, and P's best response is not to believe. Unfortunately for SB, the resulting Nash equilibria, (2,3) in Belief Game 1 and (2,4) in Belief Game 2, are next-worst outcomes for SB.[10]

While (2,4) is an NME in Belief Game 2, so is (3,3), which occurs when there is belief with revelation. However, these NMEs differ radically from those in Belief Game 1. Once in one of these states, one of

---

9. In a sequential-move belief game of incomplete information, Tarar (2018) shows that there are two Bayesian Nash equilibria. One coincides with (2,4) in Belief Game 1, at which there is not revelation but P believes nevertheless (as Pascal recommended). The second occurs when P, who may move first, does not believe because SB may not exist (a state of nature I do not assume to be possible); if this is the state, obviously SB cannot reveal himself. But if P does not believe, SB, if he exists, may reveal himself in equilibrium. By contrast, the players in the two Belief Games cannot infer information from each other's moves but, instead, anticipate them, given they have complete information about each other's preferences. Whereas the two NMEs in each of the Belief Games depend on where play starts, making them history-dependent, in Tarar's game the equilibria depend on the players' updating information in the course of play.

10. In the anticipation game, both players have weakly dominant strategies that yield [2,4] at $\overline{R}B$.

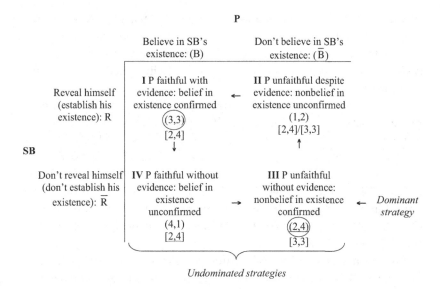

Key: (x,y) = (payoff to SB, payoff to P)
    4 = best; 3 = next-best; 2 = next-worst; 1 = worst
    Anticipation game given in brackets
    Nash equilibria underscored
    Nonmyopic equilibria (NMEs) circled
    Arrows show movement to and from NMEs

Figure 4.2
Nonmyopic equilibria in Belief Game 2 (game 49)

the players will move from it—SB from (2,4) to (1,2), P from (3,3) to
(4,1)—as shown by the arrows in the payoff matrix of figure 4.2—
whence the other player will countermove to the other NME, making
these NMEs unstable.

What, then, gives us the right to call them equilibria? They achieve
this status because they are *attractors*, each drawing in one or both
players from the two non-NMEs, (4,1) and (1,2).[11] However, once at
each of these NMEs, the players have an incentive to move to the other
NME, creating a bouncing back and forth between the two NMEs.
Consequently, I call these NMEs *boomerang NMEs*, though "ping-pong
equilibria" could serve equally well to describe the back-and-forth
movement between the NMEs in Belief Game 2.

Note that it is the NME calculations, grounded in backward
induction, that create the counterclockwise cycling. Cycling is not an

---

11. I will discuss shortly why there are two states below (1,2) in the anticipation game.

assumption—of the kind made in chapter 3—that the players move indefinitely until, possibly, a player with moving power forces termination of the cycling. Instead, the NME calculations provide a *rational basis* for the moves and countermoves that are indicated by the arrows in figure 4.2.

The fact that (2,4) and (3,3) attract as well repel in Belief Game 2 makes observable outcomes difficult to predict. Every state can occur in the process of cycling, which may make the players' choices appear random—though not entirely so, because the counterclockwise cycling specifies an order in which the states appear and disappear. Thus, before (2,4) is transformed into (3,3), the players pass through the Pareto-inferior state, (1,2); and before state (3,3) is transformed into (2,4), the players pass through (4,1), which is the best state for SB but the worst state for P.

The former state, (1,2), will surely be distressing for the players, especially SB, because despite his revelation, P continues not to believe. The latter state, (4,1), will be equally upsetting for P, because she continues to believe despite the lack of evidence, though SB will be especially pleased by P's expression of faith without revelation.

I indicate below (1,2) in the anticipation game of figure 4.2 two states, [2,4]/[3,3], separated by a slash. The state to the left of the slash is the NME that play of the game will go into if SB, the row player, moves first from (1,2), whereas the state to the right of the slash is the NME that play will go into if P, the column player, moves first from (1,2). Obviously, SB would prefer (3,3), whereas P would prefer (2,4), so each player would prefer that the other player move first from (1,2), which I call an *indeterminate state* because it remains unspecified which player will be the first to move away.[12]

I call the player who determines the order of moves the player with *order power*. In Belief Game 2, if SB has order power, he can induce (3,3), whereas P can induce (2,4) if she has order power. Manifestly, each player would prefer to exercise order power—holding out longer at (1,2) and forcing the other player to to move first from it—than for the other player to exercise it.

12. In some games the player with order power (as defined in the next paragraph) does better moving first rather than second from an indeterminate state, unlike here when each player wants to hold out longer at (1,2), forcing the other player to move first. This is reminiscent of bargaining situations, in which each player wants the other player to make the first concession.
13. In section 8.4, I analyze the conflict between Saul and God/Samuel in a game called the Deception Game.

Because cycling is counterclockwise in Belief Game 2, this player will generally be P, because it is SB who brings the players from (2,4) to (1,2), thereby giving P the next move from (1,2). This counterclockwise cycling favors P, unless (1,2) is the initial state, leaving open the player who can initiate a move from it. In this situation, it is reasonable to suppose that SB will be that player in many conflicts, in which case he can induce his next-best outcome of (3,3), which also happens to be next best for P.

In the Bible, however, it is not always God who exercises order power. For example, in the first book of Samuel, when the Israelites demand a king, "like all other nations" (1 Sam. 8:17), God grudgingly accedes to their wishes, much to His later chagrin. When the person chosen, Saul, offers what God considers to be an inferior sacrifice, however, God reasserts His control through the prophet, Samuel, who systematically undermines Saul's authority (see section 8.4).

Eventually, Saul is replaced by David, the second king of Israel, who commits numerous sins, including adultery and murder, with virtually no negative consequences; in fact, David's actions seem to receive tacit approval from God. In engineering David's accession, God and Samuel give him pretty much free rein to assert his will. When David seizes the initiative, I interpret this as David's exercise of order power, though later God does not entirely approve of some of his actions (Brams [1980] 2003, chaps. 7–8).

4.4 Conclusions

In this chapter, I have proposed new rules of play, different from those in standard game theory. Players do not begin play of a game by choosing strategies but, instead, start in some state, from which they can move and countermove. At this state, or status quo, the players make rational calculations not only about their own moves but also the moves of the other player, based on backward induction.

The results of these calculations tell them whether it is rational to stay at the initial state or move to another state and, if so, where play will terminate. I assume that the players know each other's preferences in each state, which they use to assess the consequences of their moves and countermoves.

I postulate the same set of goals, and therefore preferences, that I did for P and SB in chapter 3. I suggest that P may be of two types, at least with respect to whether she prefers belief to nonbelief as a secondary goal. In particular, P favors belief with revelation in Belief Game 1 compared to nonbelief with nonrevelation, whereas her preference is the opposite in Belief Game 2.

In both games, there are two NMEs; the one common to both games is belief with revelation. The second NME in Belief Game 1 is also associated with belief, but without revelation. In Belief Game 2, the second NME is nonbelief if there is no revelation.

While the NMEs in both games attract moves from the other two states, the NMEs in Belief Game 2 are boomerang NMEs: Once reaching either, it is rational for the players to move to the other NME. This creates a bouncing back and forth between the two NMEs, via counterclockwise cycling.

As I showed in chapter 3, this is the direction in which Belief Game 2, with one impediment, cycles. But in this chapter, I assumed that the possible moves and countermoves of players cannot proceed beyond one complete cycle, whereas in chapter 3 I assumed cycling can go on indefinitely. It terminates, however, if one player possesses moving power—and so can outlast the other player—and chooses to force the other player to stop, or both players prefer a state that only one player can induce.

The two NMEs in both Belief Games create uncertainty about what outcome will be selected. This problem is more serious in Belief Game 2, because when players look ahead, it is rational to oscillate between the two NMEs.

Clearly, the instability of the NMEs, especially in Belief Game 2, highlights the problem that P faces in deciding about the existence of SB. Although the Bible suggests that God's choices are neither arbitrary nor random—He has purposes He wants to fulfill—in games like Belief Game 2, a pattern of behavior may be difficult to discern. In particular, SB does not always get his way, as we will see in several of the later Bible stories. Thereby he may be unable to distinguish himself from P, raising questions about whether belief in SB is rational.

5   Paradoxes of Prediction

5.1   Introduction

In chapters 2, 3, and 4, I analyzed the difficulty of P deciding whether to believe or not believe in SB, whether her choice was (i) a decision in the face of uncertainty about SB's existence or (ii) as a player in a game in which SB also makes choices. In this chapter, I do not ask whether P should believe in SB but instead raise a different question: In a decision, which later becomes two different games, what choice should P make when she knows that SB can almost surely predict her choice?[1] It turns out that it is rational to make one choice if she chooses her dominant strategy, another if she wishes to maximize her expected monetary payoff, taking into account SB's predictive power.

This problem of choice was formulated by a physicist, William A. Newcomb, in 1960, elucidated by a philosopher, Robert Nozick (1969), and popularized by a writer, Martin Gardner (1973). Many scholars (mostly philosophers) have pondered this question, but it seems that nobody, in scores of articles that have been written about it, has proposed a generally accepted solution, which is why it is still considered by many as an unresolved paradox.

I will discuss one resolution that I believe clarifies certain issues. But then I will reformulate Newcomb's problem as two games in which SB and P are players with preferences. As in Newcomb's problem, SB can make predictions about P's choice in the games, the accuracy of which is known to both players.

P has different goals in the two games I postulate: (i) to maximize her monetary payoff or (ii) to demonstrate her free will by upsetting SB's prediction. These different goals lead to two different outcomes.

---

1. This chapter, except for section 5.5, is adapted from Brams (1975; 1976, chap. 6).

In the case of (i), the accuracy of SB's prediction affects the outcome, but in the case of (ii) it does not.

I next analyze a game, Chicken (game 57 in the appendix), in which SB is *omniscient*—he can make perfect predictions about what strategy P will choose. If P is aware that SB is omniscient, then there is a *paradox of omniscience*: P can induce her best outcome, whereas SB suffers his next-worst outcome. In five other 2 × 2 games of conflict besides Chicken, this paradox afflicts the omniscient player, who does worse than the nonomniscient player in terms of the players' rankings of outcomes.

I conclude that omniscience or near omniscience may be a curse as well as a blessing. By sometimes preventing SB from attaining his desired goals, it makes the task of deciding whether he is truly superior problematic.

5.2   Newcomb's Problem

Imagine the following situation. SB presents P with two boxes, B1 and B2. B1 contains $1,000; B2 contains either $1,000,000 or nothing, but you do not know which. P has a choice between two possible actions:

1. Take both boxes.
2. Take B2 alone.

Now what is in B2 depends on what action SB predicted P would take beforehand. If he predicted P would (1) take what is in both boxes— or would randomize her choice between the two actions—he put $0 in B2; if he predicted P would (2) take only what is in B2, he put $1,000,000 in B2.

Hence, P is rewarded for taking only what is in B2—provided SB predicted this choice—though P has some chance of getting even more ($1,001,000) if she takes what is in in both boxes and SB incorrectly predicted that she would take only what is in B2. On the other hand, P does much less well ($1,000) if she takes what is in both boxes—and SB predicted this action—and worst ($0) if she takes what is in B2 and SB incorrectly predicted that she would take what is in both boxes.

These payoffs are summarized in the payoff matrix of figure 5.1. Note that SB's strategies given in figure 5.1 are predictions, not what he puts in B2. But we could as well define his two strategies to be "put $1,000,000 in B2" and "put $0 in B2," because these actions are in one-to-one correspondence with SB's predictions about what P will take.

SB

		Predicts P takes B2	Predicts P takes both boxes
	Take B2	$1,000,000	$0
P			
	Take both boxes	$1,001,000	$1,000

Figure 5.1
Payoff matrix of Newcomb's problem

Therefore, it does not matter whether we consider SB's strategies to be predictions or actions. But because SB's predictions precede his actions, they are perhaps the more basic indicator of his behavior.

From a decision-theoretic perspective, what does matter is that SB's strategies are not the free choices we assumed earlier of players in a game. They are dictated by what he predicts P will do; at the same time, they are not states of nature that arise according to some probability distribution, as we assumed of decisions in chapter 2. What SB chooses depends, probabilistically, on what P chooses.

In section 5.3, we offer a different perspective. The games we define depend on both P's and SB's preferences and the probability that SB's predictions are correct. Thereby, P's and SB's rational choices are inextricably linked.

On first blush, it would appear, Newcomb's problem does not present P with a problem of choice. Her second strategy—take both boxes—strongly dominates her first strategy—take only B2—because whatever SB predicts, P's payoffs are greater than those associated with her first strategy. Thus, P should always take both boxes, which also assures her of at least $1,000, as contrasted with a minimum of $0 if she chooses her first strategy.

This choice is complicated, however, by P's knowledge of the past performance of SB, who is assumed to be superior precisely because his predictions have always been correct in the past. Although P does not know what SB's prediction is in the present situation, it will, she believes, almost surely be correct. Thus, if she chooses her dominant strategy of taking both boxes, SB will almost certainly have anticipated her action and left B2 empty. Hence, P will receive only $1,000 from her choice of both boxes.

On the other hand, if P chooses her first strategy and takes only B2, SB, expecting this, will almost surely have put $1,000,000 in B2. This would seem a strong argument for P to choose this strategy, despite the dominance of her second strategy. This argument is based on the principle of maximizing her expected payoff—associated with money in this case—which, for each of P's strategies, is the sum of her payoff at each outcome times the probability that it occurs (see section 2.2).

Assume, for purposes of illustration, that although P has near-perfect confidence in the predictions of SB, she conservatively estimates that the probability p of his being correct is only 0.9. Then the expected payoff, $EP(p)$, of P's first strategy (take only B2) is

$$EP_{B2}(0.9) = (\$1,000,000)(0.9) + (\$0)(0.1) = \$900,000,$$

whereas the expected payoff of her second strategy (take both boxes) is

$$EP_{Both}(0.9) = (\$1,001,000)(0.1) + (\$1,000)(0.9) = \$101,000.$$

Consequently, for P to maximize her expected payoff, she should take only B2. (In fact, the probability that SB is correct need only be $p >$ 0.5005 in this example to make her expected payoff from her first strategy exceed that from her second strategy.)

This conflict between the *dominance principle*, which prescribes taking both boxes, and the *expected-payoff principle*, which prescribes taking only B2, is the heart of the paradox. Although each principle can be supported by very plausible arguments, the choices that each prescribes will in general be in conflict, given SB's excellent powers of prediction.

This paradox is not the product of any hidden or suppressed assumptions. P is assumed fully to understand the choice situation, SB knows that P understands, and so on. Furthermore, no kind of "backwards causality" is assumed to be at work, whereby P's present choice influences SB's past predictions. SB is assumed to have made a prediction— say, a week before P makes her choice—and put either $1,000,000 in B2 or nothing. The money is there or it is not there, and nothing that P thinks or does can subsequently change this fact.

What will P choose? Nozick (1969, 117) reports the following:

I have put this problem to a large number of people, both friends and students in class. To almost everyone it is perfectly clear and obvious what should be done. The difficulty is that these people seem to divide almost evenly on the problem, with large numbers thinking that the opposing half is just being silly.

Although respondents to Gardner's first *Scientific American* article favored the expected-payoff principle by better than two to one, Nozick concludes his reply to the respondents by saying that "the [148] letters do not, in my opinion, lay the problem to rest" (Gardner 1974, 108).

5.3   Which Principle, and Is There a Conflict?

In Newcomb's problem, there is an obvious asymmetry between the abilities of the two players. SB is a phenomenally good predictor, but P possesses no such superior ability. Furthermore, their choices are not symmetrical—SB chooses first (predicting, and then putting or not putting $1,000,000 in B2).

In fact, however, this would appear to give SB neither an advantage nor a disadvantage, because his choice (based on his prediction) is not communicated to P. Thus, we can just as well assume that the two players make simultaneous choices, unbeknownst to each other; the essential nature of their situation is unchanged. What makes it unusual, however, is that SB's choice is governed by the prior prediction he makes, and P knows this.

In the Hebrew Bible, God helps those He favors not just with monetary rewards but in other ways as well. For example, after being severely tested, Job is able to restart his life with a new family and great riches provided by God (see section 7.4). More generally, the rewards God bestows and the penalties He inflicts on human players usually come after they demonstrate, in His eyes, good or bad behavior.

Occasionally, however, God, like SB in Newcomb's problem, acts according to His predictions of events that have not yet unfolded. As a case in point, when God tells Abraham He will spare Sodom and Gomorrah from destruction if Abraham is able to find 50, and subsequently 45, then 40, then 30, then 20, and finally only 10 righteous people in these cities (Gen. 18:26–32), it appears that He does so because He has predicted that Abraham will fail.

In fact, Abraham does fail. Then, while God allows Abraham's nephew, Lot, and his wife and daughters to escape, Lot's wife, disobeying an angel's command, is turned into a pillar of salt when she looks back at the destruction of the cities as a "sulfurous fire" from heaven rained down upon them (Gen. 19:15–26).

In Newcomb's problem, P's knowledge of SB's predictive ability evidently influences P's choice of either B2 or both boxes. Presumably, P will take only B2 if she believes SB has anticipated her choice of just this box, which is congruent with the expected-payoff principle; but she will take both boxes if she is not so sure, which is congruent with the dominance principle. As I will argue in section 5.4, choosing both boxes might also reflect P's desire to assert her free will: By choosing both boxes, she demonstrates that she will not be seduced by the probable monetary reward of choosing B2 alone, thereby showing that she is her own person and can exercise free will.

Is there any solution to Newcomb's problem that resolves the apparent inconsistency between the expected-payoff principle and the dominance principle? John A. Ferejohn (1975) has shown that if Newcomb's problem is reformulated as a different decision-theoretic problem, the apparent inconsistency between the two principles disappears.

Recall from chapter 2 that in a model of decision-making under uncertainty, the action that a player takes (in Pascal's wager, P chooses to believe or not believe in God) does not lead to a particular outcome with certainty but to a set of possible outcomes with nonzero probabilities of occurrence. If we conceptualize Newcomb's problem as a decision by P in the face of uncertainty, then instead of viewing SB as making predictions and choosing accordingly, P may see SB as making either a correct or an incorrect prediction about what she will choose (see figure 5.2). Note that P's very best ($1,001,000) and very worst ($0) outcomes occur when SB's prediction is incorrect; the intermediate outcomes ($1,000,000) and $1,000) occur when SB's prediction is correct.

Recall that P's two best outcomes in the payoff matrix of figure 5.1 ($1,000,000 and $1,001,000) are both associated with SB's predicting that P will take only B2 (first column of figure 5.1). In the decision-theoretic matrix of figure 5.2, by contrast, these outcomes are along the diagonal; each is associated with a different state of nature, which is a correct or an incorrect prediction on the part of SB.

	State of nature	
	SB's prediction correct	SB's prediction incorrect
P Take B2	$1,000,000	$0
Take both boxes	$1,000	$1,001,000

Figure 5.2
Newcomb's problem as a decision

Now P's best choice depends on what state of nature obtains in this decision-theoretic representation: If SB is correct, P should take B2; if SB is incorrect, she should take both boxes. Notice that neither of P's two strategies dominates the other.

Because P does not have a dominant strategy in this decision-theoretic representation, there is no longer a conflict between the dominance principle and the expected-payoff principle.[2] Whether P should take B2 or both boxes depends on what choice maximizes her expected payoff. If the probability p that SB is correct is greater than 0.5005, then P should take B2; if this probability is less than 0.5005, P should take both boxes; and if this probability is exactly 0.5005, P will be indifferent between her two choices.[3]

How persuasive is this resolution of Newcomb's problem? If P believes that SB, acting mechanically according to his prediction, has no control over which state of nature obtains in figure 5.2, then SB is not a player in a game. In this case, Newcomb's problem, as reformulated, is still decision-theoretic.

To be sure, the probabilities of being in each state are not specified in Newcomb's problem, so the decision-theoretic representation in figure 5.2 does not answer the question of whether P should take only B2 or both boxes. However, this representation does demonstrate that there is no necessary conflict between the dominance principle and the expected-payoff principle.

On the other hand, if you believe that SB has some control over which state of nature obtains—which is a question quite different from whether he can predict P's choice (which he almost surely can)—then he is not an entirely passive state of nature, at least with respect to being correct. In this case, a decision-theoretic representation is not appropriate, because states of nature do not occur by chance (with or without known probabilities),[4] as indicated in chapter 2 (see note 3 of that chapter).

---

2. Gilboa (2009, 113–115) modifies Newcomb's problem, under the assumption that if P has free will, SB cannot be a nearly perfect predictor of her choices. He proposes states of nature that also show there need not be a conflict between the dominance principle and the expected-payoff principle.

3. In decision theory, if the probability that an event occurs is not known, this is considered a decision under uncertainty, whereas if the probability is known, it is considered a decision under risk.

4. This lack of independence also can be problem in a game when one player controls another player's strategy choice (see section 9.3).

However, there is nothing in the original statement of Newcomb's problem to indicate that SB's choices are anything but mechanical—that is, the correctness of his prediction about P's choice is not assumed to depend in any way on P's choice. Or, to put it another way, though P is assumed to exercise free will with respect to the action she takes, SB exercises no free will with respect to what he puts in B2; his "choice" is dictated solely by his prediction.

The fact that SB's prediction is assumed to be almost surely correct would seem to imply that P is indeed playing a game against nature whose two states—SB's prediction is correct or incorrect—are tied to what P does but do not affect the probability that SB is correct. Given this interpretation of Newcomb's problem, then Ferejohn's decision-theoretic reformulation of the problem would appear to resolve the presumed conflict between the dominance principle and the expected-payoff principle.

Because any partition of the states of nature is arbitrary, however, it is not clear that the figure 5.2 partition is superior to the figure 5.1 partition, except in resolving the conflict between the dominance principle and expected-utility principle. An argument in favor of the figure 5.1 representation is that because SB's predictions in figure 5.1 precede knowledge of their correctness in figure 5.2, the figure 5.1 partition offers a more fundamental representation of the decision problem than the figure 5.2 representation.

We next ask what consequences the predictive ability assumed of SB would have if (i) SB, as well as P, has preferences—his goal, presumably, is to make correct predictions—for the four possible outcomes and (ii) the quality of SB's prediction is known. Then both P and SB can, based on their preferences, make rational choices in a game.

This line of inquiry echoes our earlier reformulation of Pascal's wager as two Belief Games in chapters 3 and 4, in which *both* P and SB each can make choices. But we add two new elements: (i) the players know the probability that SB's prediction is correct and (ii) P's preference may not be to maximize her monetary reward but, instead, to show her independence—specifically, by not doing what SB predicts—which is quite a different goal from that assumed in Newcomb's problem.

5.4   Two Prediction Games

As with the two Belief Games in chapter 3, I specify two games to represent Newcomb's problem when it is viewed as a game. In

Prediction Game 1 (game 48 in the appendix), I assume that the (i) primary and (ii) secondary goals of the players are the following:[5]

SB: (i) wants to make a correct prediction of P's choice; (ii) prefers to put $1,000,000 in B2.

P: (i) wants SB to predict B2 and, hence, put $1,000,000 in it; (ii) prefers to choose both boxes.

The payoff matrix that these goals imply is shown in figure 5.3. SB's two best outcomes (4 and 3) are associated with his correct predictions—along the main diagonal—and his two worst outcomes (2 and 1) are associated with his incorrect predictions—along the off-diagonal. Consistent with SB's secondary goal, SB's preferred outcomes in each case (4 when he is correct, 2 when he is incorrect) are associated with his predicting that P will choose B2 (first column) and, therefore, putting $1,000,000 in this box.

I assume that P wants SB to put $1,000,000 in B2 (primary goal), because she wants to maximize the money she receives. This goal is best achieved when SB predicts P will choose B2 (first column). P's two worst outcomes occur when SB predicts P will choose both boxes and puts nothing in B2 (second column). Given that SB has made his choice and cannot change it, P would prefer to choose both boxes (secondary goal), because this gives her a minimum of $1,000 and, possibly, $1,001,000.

These monetary payoffs for P are shown below both players' ordinal payoffs, or ranks, of the four outcomes, and the anticipation game payoffs, in figure 5.3. Note that P's monetary payoffs, from $1,001,000 to $0, correspond to her rankings from 4 to 1 in Prediction Game 1 (but not to P's ordinal payoffs in the anticipation game).

It turns out that Prediction Game 1 has the same ordinal payoffs (i.e., ranks) as Belief Game 1 (figures 3.1 and 4.1), except the roles of P and SB are now reversed.[6] P has a dominant strategy of choosing both

5. Unlike Prediction Game 1, I assume in Brams (1975; 1976, chap. 6) that both players have the preferences of P, which yields Prisoners' Dilemma (game 32 in the appendix). They can escape the dilemma in this game if one player, as the leader, announces a policy of "conditional cooperation" (I'll cooperate if I predict you will cooperate; otherwise, I won't), and the follower rationally responds by cooperating if the leader's probability of making a correct prediction is sufficiently high. In Prediction Game 1, as I will show later in this section, P can be induced to choose her dominated strategy if SB's probability of making a correct prediction is sufficiently high.
6. Thus, P's payoffs correspond to SB's payoffs, and SB's to P's, in Belief Game 1.

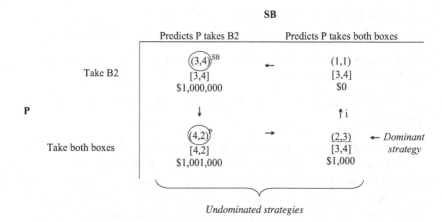

Figure 5.3
Prediction Game 1 (game 48)

boxes, which gives the Nash equilibrium of (2,3), which is Pareto-inferior to (3,4) when SB predicts the choice of both boxes.

As I showed for Belief Game 1 (figure 4.1), Prediction Game 1 has two NMEs, (3,4) and (4,2)—each reachable from the states with these payoffs that are shown in brackets in the anticipation game. P and SB, respectively, can also induce these outcomes with moving power, which is effective. Because Prediction Game 1 is moderately cyclical in a counterclockwise direction, all outcomes except (1,1) have some claim to being implemented, under different conditions, in Prediction Game 1.

But what if one player (SB in Newcomb's problem) can almost surely anticipate the other player's (P's) strategy choice? What then is the likely outcome? Assume, as before, that the probability that SB's prediction is correct is p, and both players know this. Assume, for the purpose of illustrating the next calculation, that the ranks in figure 5.3 are cardinal utilities, or payoffs that denote the numerical value of the outcomes to each player. Given SB's strategy is mechanical, based on his prediction, what is P's rational strategy choice?

As before, define P's *expected payoff* from choosing a strategy as the sum of her payoff at each outcome associated with this strategy times the probability that it occurs. Then P's expected payoff from choosing B2 is

$$EP_{B2}(p) = 3p + 1(1 - p),\tag{5.1}$$

because she obtains a payoff of 3 (in Newcomb's problem, her next-best payoff of \$1,000,000) when SB correctly predicts her choice with probability p and puts \$1,000,000 in B2, and a payoff of 1 (\$0) when SB incorrectly predicts her choice and puts no money in B2.

By the same token, P's expected payoff from choosing both boxes is

$$EP_{Both}(p) = 4(1 - p) + 2p.\tag{5.2}$$

It is easy to show that (5.1) exceeds (5.2) when $p > 3/4$. If we substitute the monetary values in Newcomb's problem for P's ranks of 4 (\$1,001,000), 3 (\$1,000,000), 2 (\$1,000), and 1 (0), then

$$EP_{B2}(p) = (1,001,100)p + 0(1 - p),\tag{5.3}$$

$$EP_{Both}(p) = (1,001,000)(1 - p) + (1,000)p,\tag{5.4}$$

and (5.3) exceeds (5.4) when $p > (1,001,000)/(2,000,000) = 0.5005$, or slightly more than $1/2$.

This is the same probability that we showed in section 5.2 makes it profitable for P to choose B2, based on the expected-payoff principle. Thus, formulating Newcomb's problem as a game is consistent with viewing it as a decision, in which SB is a state of nature, albeit one whose predictions are tied to his predictions about P's choice.

But there is a significant difference between the players' choices in Newcomb's problem when it is viewed as a game rather than as a decision by P, who is assumed to take into account SB's prediction capability. In the decision, P knows that SB will put \$1,000,000 in B2 with probability p (when he correctly detects P will choose B2). If $p < 1$, however, the outcome for P in the decision will not always be \$1,000,000 but sometimes be \$0 when SB incorrectly predicts P's choice of B2 (with probability $1 - p$); in that case, SB will put nothing in B2.

In the game, SB is assumed to make his own rational calculation. He knows that if p is greater than some threshold value (3/4 or 0.5005 in the above example), P will choose B2 to maximize her expected payoff.

Hence, one might think that SB should *always* put $1,000,000 in B2. However, if SB does this, P should choose both boxes to maximize her payoff, obtaining $1,001,000 instead of $1,000,000. Because SB cannot rest assured that P will always choose B2 if he always puts $1,000,000 in B2, he is well advised to rely on his imperfect detector, as in the decision-theoretic formulation.

But what if P's primary goal is not to maximize her monetary reward but, instead, to show her free will by attempting to falsify SB's predictions about what she will do. This goal is very different from that in Newcomb's problem, which Isaac Asimov justified with a provocative contrarian argument:

> It is perfectly clear to me that any human being worthy of being considered a human being (including most certainly myself) would prefer free will. ... If God has muffed and left a million dollars in the box [B2], then not only will you have gained that million but far more important you will have demonstrated God's nonomniscience. If you take only the second box [B2], however, you get your damned million and not only are you a slave but also you have demonstrated your willingness to be a slave for that million and you are not someone I recognize as human. (quoted in Nozick 1997, 81)

In other words, if you are human, you want to do your best to defy SB by falsifying his prediction, even if this costs you a monetary reward; but if that does not work, you want to show SB that you are not a slave who can be seduced into choosing B2.

In Prediction Game 2 (game 42 in the appendix), I assume SB's primary and secondary goals remain the same as in Prediction Game 1. But P's (i) primary and (ii) secondary goals change to those that follow, which gives rise to the game depicted in figure 5.4:

P: (i) wants SB's prediction to be incorrect; (ii) prefers to choose both boxes.

Prediction Game 2 has no Nash equilibrium: From every state, one player has an incentive to move in a counterclockwise direction, making the game strongly cyclic (and thus highly unstable).[7] However, it has two NMEs, (4,2) and (2,3), that P and SB, respectively, can induce with moving power or, alternatively, from the states indicated in the antici-

---

7. If the payoffs to players were cardinal utilities rather than ranks, then this game—and Chicken, which I analyze in section 5.5—have a Nash equilibrium in "mixed strategies," whereby players randomize their choices of strategies according to some probability distribution. But such strategies are difficult to interpret in the context of one-shot play, so I limit consideration to single choices, or "pure strategies," in applying standard game theory.

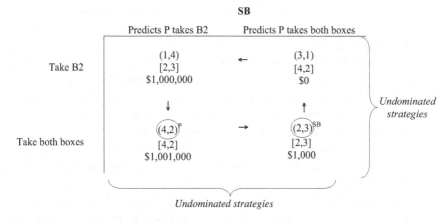

Figure 5.4
Prediction Game 2 (game 42)

pation game. But what is most relevant here is how P's contrary primary and secondary goals in Prediction Game 2—to upset SB's prediction, or at least not choose B2—affect her rational choice of a strategy.

If the ranks are utilities, as we assumed in Prediction Game 1, P's expected payoff from choosing B2 is

$$EP_{B2}(p) = 1p + 3(1 - p),\tag{5.5}$$

and her expected payoff from choosing both boxes is

$$EP_{Both}(p) = 4(1 - p) + 2p.\tag{5.6}$$

For no value of p is it rational for P to choose B2, because $4{-}2p$ from (5.6) is always greater than $3{-}2p$ from (5.5). Thus, P maximizes her expected payoff by *always* choosing both boxes, despite the fact that neither of her strategies is dominant (unlike Prediction Game 1, in which choosing both boxes was dominant): If SB predicts B2, it is rational for P to choose both boxes; if SB predicts both boxes, it is rational for P to choose B2.[8]

---

8. For a more general analysis of this issue, applicable to all the 2×2 strict ordinal games, see Brams and Kilgour 2018.

To understand better why P will always choose both boxes, suppose $p = 1$, so SB predicts perfectly which strategy P will choose. Then P, by choosing B2, will obtain an ordinal payoff of 1, as given by (5.5), whereas her payoff will be 2, as given by (5.6), if she chooses both boxes. Likewise, if $p = 0$, P's payoff will be 3 from choosing B2 and 4 from choosing both boxes. Not only does P always do better by choosing both boxes at these extreme values of p, but she also does so for all intermediate values of p strictly between 0 and 1.

In the decision that corresponds to Prediction Game 2, SB's choices are mechanical, based on his prediction of P's choice: Put $1,000,000 in B2 if he predicts P will choose B2, nothing if he predicts that P will choose both boxes. As in Prediction Game 1, if $p < 1$, SB will sometimes incorrectly predict both boxes and put nothing in B2 when P chooses B2, which will give P a payoff of $0. But by choosing both boxes, she obtains a minimum of $1,000.

In Prediction Game 2, SB knows that it is rational for P always to choose both boxes—independent of the value of p—because choosing both is her strongly dominant strategy. Hence, he should not base his choice on his less-than-perfect prediction of P's choice. Instead, surmising that P will always choose both boxes, SB will never put $1,000,000 in B2. When both players choose their rational strategies (P chooses both boxes; SB, knowing P's rational choice of both boxes, puts nothing in B2), P will obtain $1,000, which is her next-worst outcome, and SB will obtain his next-best outcome, yielding (2,3) in Prediction Game 2.

This outcome, like all others in Prediction Game 2, is not a Nash equilibrium, but it is an NME if play starts at either (2,3) or (1,4); also, it can be induced by SB's moving power. Although P wishes to upset SB's prediction, rational play in Prediction Game 2 leads P always to choose both boxes, whatever the value of p is, and for SB to anticipate this choice and so put nothing in B2. Therefore, while P does not succeed in upsetting SB's prediction, she at least denies him the satisfaction of achieving his secondary goal of putting $1,000,000 in B2. In effect, she prefers to demonstrate her free will, accepting only a small reward ($1,000), rather than being seduced by the $1,000,000 prize.

Interpreted as games, the outcomes in Newcomb's problem depend very much on the goals that the players want to achieve. In Prediction Game 1, P will choose B2 if p is sufficiently high, and SB, knowing this, will put $1,000,000 in B2, leading to (3,4) with probability p. In Prediction Game 2, on the other hand, P will always choose both boxes, independent of p; SB will anticipate this choice and put nothing in B2, rendering (2,3) the outcome.

The decision that corresponds to Prediction Game 1 will, based on the expected-payoff principle, produce the same outcome—P will choose B2 if p is sufficiently high, obtaining $1,000,000 with probability p and nothing with probability $1 - p$. It is hard to say what decision corresponds to Prediction Game 2, unless new utilities are attached to the different outcome that reflect P's goals. Conceivably, if SB knows that monetary payoffs do not matter to P as much as contradicting his predictions, SB might randomize his choice of what he puts in B2, rendering P's attempt to make a contrary choice right, on average, half the time. This is in line with a player's choice of mixed strategies in a game without a pure-strategy Nash equilibrium (see note 7).

The two prediction games do not exhaust possible interpretations of Newcomb's problem as a game, but they do suggest that a game—in which SB as well as P has preferences—may lead to determinate choices. While P's rational choice of B2 in Prediction Game 1 depends on p's being sufficiently high, her rational choice of both boxes in Prediction Game 2 is independent of p.

Thus, instead of having to make imperfect predictions, SB need only anticipate P's certain rational choice (both boxes in Prediction Game 2) and then respond accordingly. Although P fails to upset SB's prediction in this game, she does succeed in frustrating SB's secondary goal of bestowing $1,000,000 on her for choosing B2. Instead, P receives the paltry sum of $1,000 for defying SB by choosing both boxes, but she gains the satisfaction of preventing SB from achieving both his goals.

5.5 The Paradox of Omniscience[9]

The outcomes in Prediction Games 1 and 2, (3,4) with probability p and (2,3) with certainty, respectively, each favor SB in terms of their comparative rankings by (P, SB).[10] Are there any prediction games in which P obtains a more-preferred outcome than SB, so SB's powers of prediction backfire?

9. This section is adapted from Brams 1994, chap. 6, and Brams 2011, chap. 9.
10. Note that rankings avoid interpersonal comparisons of utilities, because they make the standard of comparison not how well one does vis-à-vis an opponent but how well one does vis-à-vis oneself. Thus, for example, P obtains her best outcome (4), whereas P obtains her next-best outcome (3), with probability p in Prediction Game 1, which is the basis for saying that SB does better than P.

Assume that SB's superiority rests not only on being able to predict P's strategy choice with probability $p > 1/2$ but also that $p = 1$. That is, SB is *omniscient*—he can predict P's strategy choice perfectly. While P has no such ability, assume that she is aware of SB's omniscience, perhaps because SB revealed himself in one of the Belief Games.

Consider the game shown in figure 5.5 (game 57 in the appendix). It is *symmetrical*, because each player's payoff below the main diagonal, (4,2), is a mirror image of his or her payoff, (2,4), above the diagonal. Furthermore, the players' payoff rankings along the diagonal, (3,3) and (1,1), are the same.

As shown in figure 5.5, this game, known as Chicken, has three NMEs. It has been used to model a variety of confrontation situations, such as the conflict between the two superpowers—the United States and the Soviet Union—during the Cuban missile crisis of October 1962 (for a critical assessment, see Brams [1994, 2011]).

The story commonly told to illustrate Chicken is the following: Two car drivers race toward each other on a narrow road. Each has the

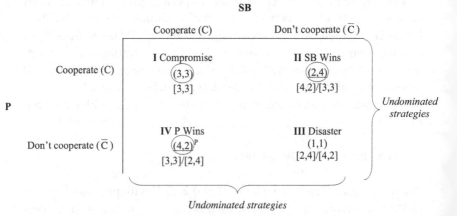

Key: (x,y) = (payoff to SB, payoff to P)
    4 = best; 3 = next-best; 2 = next-worst; 1 = worst
    Nash equilibria underscored
    Nonmyopic equilibria (NMEs) circled
    Anticipation game given in brackets
    Superscript P indicates outcome P can induce if SB is omniscient and P is aware of
        SB's omniscience.

Figure 5.5
Chicken (game 57): A confrontation game

choice of swerving, and avoiding a head-on collision (its cooperative strategy, C), or continuing on the collision course ($\overline{C}$). Each player would

- most prefer (4) that the other player "chicken out" by choosing C, when it chooses $\overline{C}$, at $\overline{C}$C or $\overline{C}\overline{C}$.
- second-most prefer (3) to chicken out by choosing C, when the other player also does at CC, and so be mildly disgraced;
- third-most prefer (2) to be the sole player to chicken out, when it chooses C at $\overline{C}$C or $\overline{C}$C, and so be severely disgraced;
- least prefer (1) not to chicken out by choosing $\overline{C}$, when the other player also does at $\overline{C}\overline{C}$, so both die in a head-on collision.

A variant of this story is that both drivers race toward a cliff, where each tries to be the last to jump out from his or her car before it sails over the cliff.

As with earlier games, we can define Chicken in terms of the (i) primary and (ii) secondary goals of each player:

(i) prefers that the other player choose C;
(ii) prefers to choose C if the other player chooses $\overline{C}$; prefers to choose $\overline{C}$ if the other player chooses C.

A *paradox of omniscience* occurs when it is better to be the nonomniscient player P than the omniscient player SB, given that P is aware of SB's omniscience.

In Chicken, consider the consequences of P's choices:

- If she chooses C, SB will predict her choice and choose $\overline{C}$, giving (2,4).
- If she chooses $\overline{C}$, SB will predict her choice and choose C, giving (4,2).

Because P prefers (4,2) to (2,4), P will choose $\overline{C}$, forcing SB to back down, receiving only his next-worst payoff (2), whereas P obtains her best payoff (4). Since P cannot improve on her payoff, whereas SB most definitely can, omniscience clearly hurts SB in Chicken when P is aware of it and acts to maximize her own payoff.

Besides Chicken, there are five other 2 × 2 games (games 51–55 in the appendix) in which there is a paradox of omniscience (Brams 1994, 2011). Neither player in any of these games has a dominant strategy, and each game has two Nash equilibria, which are also its NMEs. These are also games in which order power in applicable at the states at which a player can induce one of two NMEs.[11]

To have omniscience in a game is equivalent to a player's moving second, giving him or her the opportunity to observe the prior choice of an opponent and respond to it. Omniscience eliminates the necessity of having to move second, because one can act on one's certain knowledge of what an opponent will do before he or she makes a choice. In Chicken, in particular, omniscience enables SB to anticipate P's choice of C or $\overline{C}$ without actually having observed it beforehand. But P, aware that SB will rationally respond to his prediction of her choice, can capitalize on this fact in games vulnerable to the paradox of omniscience.

The resolution that TOM provides to the paradox of omniscience is that it enables a player to depart from a state that the paradox initially induces. In Chicken, for example, SB can depart from the state induced by the paradox, (4,2), to the NME of (3,3) if it has order power. Without such power, however, SB's omniscience, and P's awareness of it, put SB in a clearly inferior position.

5.6   Conclusions

I have surveyed different paradoxes of prediction, beginning with Newcomb's problem, in which there is a clash between the dominance principle and the expected-payoff principle. According to the former, P should take both boxes, but according to the latter, P should take only B2. As Ferejohn (1975) showed, however, there is no conflict if Newcomb's problem is posed as a decision in which the states of nature are that SB is either correct or incorrect.

This resolution does not take into account that SB, like God in the Bible, may have preferences and be a player, like P, in a game. In Prediction Game 1, P wants SB to predict B2 and, hence, put $1,000,000 in it; it is rational for P to choose B2 if p is sufficiently high (slightly greater than 1/2 for the monetary values in Newcomb's problem). Given that SB anticipates P's choice of B2, he will put $1,000,000 in B2, but he

---

11. If either player has "threat power"—which is a third kind of power, besides moving (see chapter 3) and order (see section 4.3) power, that will be defined and illustrated in section 6.3—then that player can induce a most-preferred outcome ((4,2) for P, (2,4) for SB). I defer discussion of this kind of power, because the focus of this section is on how information in a game (SB's omniscience and P's awareness of it) affects the choice of an outcome in Chicken and related games.

cannot always do this, because then P will have an incentive to choose both boxes and obtain $1,001,000, which is SB's next-worst outcome.

Prediction Game 2 echoes Asimov's contention that P, if human, will desire to demonstrate her free will by acting contrary to SB's prediction. Given that SB surmises that it is always rational for P to take both boxes—independent of his probability of making a correct prediction—he will put nothing in B2. The result is at best a middling outcome for both players (SB does not reward P with $1,000,000; P gets only $1,000), so neither player obtains what he or she most desires.

If SB's predictive powers are perfect, making him omniscient, he can be hurt if P is aware of his omniscience. In Chicken, for example, P, anticipating SB's response to each of her strategies, can induce her best outcome, which is next worst for SB. To be sure, SB can move subsequently from this initial state, according to TOM, but it nevertheless seems strange that P can seize the advantage at the beginning of play.

In the Bible, God's ability to predict the behavior of human players, including His chosen people, the Israelites, is far from perfect, in part because humans are endowed with free will (see section 1.3). Furthermore, some human players seem able to anticipate what God desires and act accordingly. For example, after God commands Abraham to sacrifice his son, Isaac, Abraham appears to understand that his faith is only being tested—this is, he will not, in fact, have to carry out the sacrifice. Although he prepares to carry it out, he is prevented from doing so (see section 7.2).

I conclude that information, and predictions based on it by SB, not only may be imperfect but, if they are not, can be a double-edged sword. In particular, SB's ability to predict the future—when known by a human protagonist—may undermine his ability to achieve preferred outcomes, which calls into question his supposed superiority.

6 The Constraint and Temptation Games

6.1 Introduction

In chapter 1, I argued that God had good reason to endow His human subjects with free will. Still, God did not give Adam and Eve unrestricted access to knowledge, presumably to prevent them from becoming godlike and, consequently, able to challenge His supremacy.

This point is made most forcefully after the creation, when Adam and Eve enter the garden of Eden. They had already been warned by God against eating from the tree of knowledge of good and bad; in fact, they were told they would die if they violated His edict (see section 1.3). But then, after eating fruit from this tree, God reneged on his threat and inflicted a milder punishment, possibly jeopardizing the credibility of His future threats but, as I will argue, showing His mercy under trying circumstances.

Why did Adam and Eve ignore God's warning, and why did God retreat from His threat to kill them if they did? Because Eve was inveigled by a cunning serpent, who belittled God's warning, she could claim she was unwittingly enticed to eat the forbidden fruit, which she subsequently shared with Adam. The serpent's enticement, perhaps, made the pair less culpable, providing God with an excuse for not killing them.

Extrapolating from this story, I seek to explain why, in a game I call the Constraint Game, SB would impose constraints on P that he could anticipate would be abrogated. It helped that there was an agent of evil in the Adam-and-Eve story (i.e., the serpent) to entice P. But in the Constraint Game, I do not posit such an agent in order to capture situations in which P, in the absence of an evil character to egg her on, nonetheless bridles under the constraints that SB imposes.

I model the Constraint Game as a game of total conflict between SB and P, at whose Nash equilibrium SB imposes constraints and P violates them. Unlike the two Belief Games, this game has only one NME, which coincides with the Nash equilibrium. The game is cyclic in a counterclockwise direction, but in it moving power (see section 3.2) is *ineffective*—perversely, it helps the player who does *not* possess it attain a better outcome than if he or she possessed it. Although conflicts between God and human players abound in the Bible, few put the players so much at odds as does the Constraint Game.

In the second game I analyze in this chapter, the Temptation Game, I begin by describing the partial conflict between the serpent and Eve in the Bible. In reinterpreting it as a game between SB and P, SB replaces the serpent and tests P's ability to resist temptation.[1]

The serpent, it seems, is inserted in the biblical story for the same reason that Satan is inserted into the story of Job—as an evil counterpart of God, meant to do His "dirty work." In the case of Job, whose plight I analyze in section 7.4, God removes the protection that a righteous man, Job, had received and gives Satan permission to punish him unremittingly.

Both the serpent and Satan serve the purpose of introducing a surrogate of evil, as a player, into the lives of human characters. How well are humans able to hold up to stress or torment designed to shake their belief in God? In the case of Eve, the stress is created by temptation, whereas the torment in the story of Job is created by the horrendous punishment that Satan inflicts upon him.

The temptation of the serpent proves irresistible to Eve in the Bible story, though she knows that succumbing to it could be costly. In the Temptation Game, SB has a dominant strategy of tempting P, and P—even without a serpent—has a dominant strategy of succumbing to temptation. This leads to the unique Nash equilibrium and NME in this game, which is reinforced if P uses a "compellent threat" (to be defined and illustrated).

Although the Temptation Game is cyclic, SB's moving power offers no relief, inducing a Pareto-inferior outcome that neither player desires. We ask whether SB, perhaps, desires that P succumb to temptation so that he not only can punish her but also teach the world a lesson about the consequences of defying his command.

---

1. As in the Constraint Game, I do not assume an agent, like the serpent in the Bible, tempts P. SB does so himself in the Temptation Game, which makes his preferences different from those I assumed of the serpent in the Bible (Brams [1980] 2003, chap. 2).

6.2   The Constraint Game: Adam and Eve in the Garden of Eden

SB can impose (I) or not impose ($\bar{\text{I}}$) constraints on P, and P can adhere (A) or not adhere ($\bar{\text{A}}$) to them. The primary and secondary goals of SB and P that I assume are as follows:

SB: (i) wants P to adhere to constraints; (ii) prefers not to impose constraints if P would adhere to them anyway, but to impose constraints if she would not adhere to them.

P: (i) wants not to adhere to constraints; (ii) prefers constraints be imposed if she would adhere to them anyway, but not be imposed if she would not adhere to them.

Clearly, the players' two best and two worst outcomes, based on their primary goals, are diametrically opposed: SB wants P—but P does not want—to adhere to the constraints. In addition, the players' secondary goals are also diametrically opposed: SB prefers no constraints if there is adherence, but otherwise he prefers constraints, whereas P has the opposite preferences.

The players' conflicting preferences yield the game shown in figure 6.1, which is game 25 in the appendix, wherein what is best for one player (4) is worst for the other (1), and what is next best for one (3) is next worst for the other (2). This is a *game of total conflict*, or a constant-sum game if the payoffs to the two players were cardinal utilities or numerical values that sum to the same constant at every outcome (if the ranks are assumed to be utilities, the sums of the ranks are 5 in the figure 6.1 game).[2]

But what does it mean for P not to adhere to constraints if SB does not impose them? In that situation, I assume that P is aware of SB's preferences about what constraints she might voluntarily observe. She must make a choice of how to act, not sure of whether constraints that she may anticipate will or will not be imposed.

As for SB, if he does not impose constraints, he would presumably rue the day he did not impose them, because he cannot so easily punish P retroactively for disobeying a nonexistent constraint. Although it may seem a more serious breach for P to defy an existent constraint, I rate this outcome for SB, I$\bar{\text{A}}$, better (2) than $\bar{\text{I}}\bar{\text{A}}$ (1), when P acts unrestrainedly without there being any constraints. After all, SB can more

---

2. All other games in this book are games of partial conflict, so the payoffs at every outcome do not sum to the same constant.

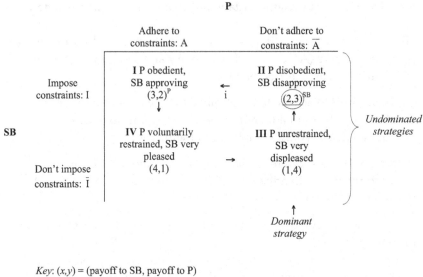

Figure 6.1
Outcome and payoff matrix of the Constraint Game (game 25)

easily exact retribution, without appearing to act arbitrarily, if he imposes constraints.

In the Bible, the payoffs in this game mimic the preferences of God and Adam and Eve. Adam and Eve's two best outcomes occur when they do not adhere to God's prohibition on eating from the tree of knowledge of good and bad; between these, they would obviously prefer not to defy God (when He does not impose constraints) than to defy Him (when He does). Their two worst outcomes occur when they adhere to God's constraints. I assume voluntary compliance, if constraints are not imposed, is least attractive to Adam and Eve, because they could just as well have chosen not to comply, given they had not been warned they would die for violating God's prohibition.

The assumption that Adam and Eve most prized their freedom from constraints requires further justification. Would they really prefer to suffer God's wrath, and the possible punishment this brings, than accept certain limits on their activities? Surely, it might be argued, God's threat of death for eating from the tree of knowledge of good and bad should act as a sufficient deterrent.

Consider, however, how the serpent, "the shrewdest of all the wild beasts that the LORD God had made" (Gen. 3:1), effectively blunted this threat when it confronted Eve. First, it asked her, disingenuously,

Did God really say: You shall not eat of any tree of the garden? (Gen. 3:1)

Note how the serpent introduced information into its question— namely, that there was (possibly) a prohibition against eating from any tree—so as to be able to deny later its alleged consequences. When Eve responded that the fruit from only one tree "in the middle of the garden" (Gen. 3.3) was forbidden, the serpent made light of this prohibition:

You are not going to die, but God knows that as soon as you eat of it your eyes will be opened and you will be like divine beings who know good and bad. (Gen. 3:4–5)

In fact, by acting naively in the beginning—not like the shrewd beast the reader is told it is—the serpent rendered itself more believable to Eve and was thereby better able to dispel her fears. But more than laying to rest Eve's fears, the serpent also offered a cogent reason for eating the forbidden fruit: It would make her divine, or godlike. This reason was reinforced by the fact that "the tree was good for eating and delight to the eyes, and ... desirable as a source of wisdom" (Gen. 3:6).

Had the serpent only ridiculed God's threat without extolling the alleged virtues of eating the forbidden fruit, I assume that its challenge to God would have been less enticing to Eve. But perceiving that an innocuous question, or just its dismissal of God's threat, would probably not suffice to tempt Eve, the serpent sought to make as airtight a case as possible for eating the forbidden fruit.

As the Bible reports, the serpent indeed was persuasive. Eve not only was attracted to the fruit and ate it herself, but she also gave some to Adam, who ate it too.

Because God's choice to impose constraints preceded Adam and Eve's choices, the proper representation of their game is not as a 2 × 2 matrix but as a 2 × 4 matrix, in which Adam and Eve's two choices are contingent on God's two choices, which would give the couple a total of four (contingent) choices (for example, choose A if God chooses I; otherwise, choose A̅). But in the Constraint Game, it is reasonable to suppose that before either SB or P chooses a strategy, both players contemplate the four possible outcomes of the 2 × 2 game and the consequences they entail.

These include possible moves away from each state, which I will model with TOM. But first consider what standard game theory prescribes in the Constraint Game. Although P has a dominant strategy of nonadherence (A̅), SB's strategies are undominated; by anticipating P's choice of her dominant strategy in a game of complete information, SB will impose constraints (I), leading to the Nash equilibrium with payoffs of (2,3).

This equilibrium is also the unique NME, which is why I do not show an anticipation game in brackets below the players' payoffs in figure 5.1, as I did for the NMEs in the two Belief Games in chapter 4. Whatever is the initial state at which play starts in the Constraint Game, play will end up at (2,3), so there is no need to repeat these anticipation-game payoffs in brackets below each of the four states.

To illustrate why the Constraint Game has only one NME, assume that play starts at (3,2). A backward-induction calculation, analogous to those made in chapter 4 for the two Belief Games, shows that SB would move from (3,2) to (4,1), P would countermove from (4,1) to (1,4), and SB would counter-countermove from (1,4) to (2,3), where play would terminate: (3,2) → (4,1) → (1,4) → (2,3).

Part of this path would also be followed if the initial state were either (4,1) or (1,4). Finally, if play started at (2,3), neither player, including SB who obtains a payoff of 2, would have an incentive to move. If SB did move to (1,4), there would be stoppage, which is obviously worse for SB, so he will not move, even though he receives only his next-worst payoff at (2,3).[3]

The arrows in figure 5.1 indicate that the game cycles in a counter-clockwise direction, with one impediment from (2,3) to (3,2), making the Constraint Game moderately cyclic (see section 3.2). Recall that a game is cyclic if no player ever departs, in the direction of the arrows, from his or her best state (4), even though, when there is an impediment, he or she does immediately worse by moving.

---

3. The Constraint Game, which is game 25 in in the appendix, is not the only 2 × 2 game of total conflict. There are two other configurations of the four payoffs (games 11 and 44) that are total-conflict games with different NMEs.

What are the consequences if one player has moving power?

• If P has moving power, she can force SB to stop at either at (3,2) or (1,4), when SB has the next move. Because SB would prefer (3,2), this is the outcome he would choose, which is next worst for P.

• If SB has moving power, he can force P to stop at either (2,3) or (4,1), when she has the next move. Because P would prefer (2,3), this is the outcome she would choose, which is next worst for SB.

Paradoxically, whichever player has moving power, he or she does worse exercising it than if the other player exercised it. In other words, the player with moving power is penalized for using it, which makes moving power *ineffective* in the Constraint Game.[4] It belies the notion that power is something to be prized; quite the contrary, SB in the Constraint Game would prefer that P exercise moving power, and vice versa.

The fact that the exercise of moving power backfires compounds the difficulty of deciding which player in the Constraint Game is SB, based on the outcomes that each can induce with moving power. Rather than relishing the exercise of this kind of power, each player would disdain it, desiring to cede its use to the other player.

On the other hand, the fact that (2,3) is both a Nash equilibrium and an NME surely makes it more likely to be chosen than the other middling outcome, (3,2). Because P is favored by (2,3), she would appear to have greater claim to superiority in the Constraint Game, suggesting a reversal of the usual player roles in this game.

Moving power is effective in the Temptation Game, as we will next see, but the situation is not much improved for SB, wherein a different kind of power comes into play that can, in principle, help P achieve her best outcome. This outcome is also the unique Nash equilibrium and NME, so again SB may be thwarted in his ability to achieve a favorable outcome.

---

4. Recall that moving power is *effective* in Belief Game 1 (see section 3.2), wherein each player does better exercising it than if the other player does. It is *irrelevant* in Belief Game 2 (see section 3.3), wherein both players benefit when P, rather than SB, exercises it. Hence, it is rational for SB to defer to SB, which I suggested earlier is paradoxical—but not so paradoxical as when *both* players would prefer that the other player exercise moving power. In most of the 36 2×2 cyclic games, moving power is either effective (in 16 games) or irrelevant (also 16), but in four cyclic games, including the Constraint Game, it is ineffective, indicating that this anomalous case is relatively rare.

6.3   The Temptation Game: Eve and the Serpent

To return to the Bible, God did not forbid the serpent from approaching Eve. Rather, just as God permitted Satan to inflict terrible pain and suffering on Job (see section 7.4), He allowed the serpent to use all its wiles to tempt Eve to eat the forbidden fruit. When Adam and Eve do partake of the fruit,

> then the eyes of both of them were opened and they perceived that they were naked; and they sewed together fig leaves and made themselves loin-cloths. (Gen. 3:7)

Previously, after Eve had been formed from one of Adam's ribs, but before her temptation by the serpent, the Bible says:

> The two of them were naked, the man and his wife, yet they felt no shame. (Gen. 2:25)

Evidently, if the eyes of Adam and Eve were opened upon eating the forbidden fruit—as the serpent had predicted they would be— their opening did not make Adam and Eve divine. Instead, their nakedness became shameful as a consequence of both their sin against God and their acquisition of knowledge of good and bad. So they covered up.

Because Eve claims that she was "duped" (Gen. 3:13) by the serpent, God punishes it, condemning it to crawl and eat dirt for the rest of its life. But Adam and Eve are not let off the hook: Eve will suffer pain in childbirth, perhaps because she instigated Adam to sin and is therefore more culpable, and Adam will rule over her as her husband. But Adam must labor hard for his food and return to the ground as dust.

After pronouncing these punishments, God made clear why the knowledge gained from eating the fruit of the tree of knowledge of good and bad posed a problem:

> Now that man has become the one of us, knowing good and bad, what if he should stretch out his hand and take also from the tree of life [which also was in the middle of the garden of Eden] and eat, and live forever! (Gen. 3:22)

In other words, man, having become divine in his ability to distinguish good from bad, could threaten God's supremacy if he also gained immortality from the tree of life. Consequently, God chose a punishment that degraded Adam's stature, but He did not condemn him to death:

So the LORD God banished him [man] from the garden of Eden, to till the soil from which he was taken. He drove the man out, and stationed east of the garden of Eden the cherubim and the fiery ever-turning sword, to guard the way to the tree of life. (Gen. 3:23–24)

Thereby the hierarchy, with God on top, was established, which presumably was God's best outcome under the circumstances.

I say "under the circumstances," because if Adam and Eve had not eaten from the tree of knowledge of good and bad, then God, it seems, would not have felt the need to banish them from the garden of Eden. Banishment became God's rational choice only after Adam and Eve came to know good and bad. If they had not acquired this knowledge by eating the forbidden fruit, then the fruit of the tree of life would not have rendered them a threat: Immortality alone—without knowledge of good and bad—would make them an uninformed, and seemingly unthreatening, presence.

It was the combination of immortality and knowledge that God considered intolerable. He took the appropriate measures to arrest the immortality of an already knowledgeable Adam and Eve and thereby ensured His own unique and privileged position.[5]

This calculation is quite straightforward and, therefore, does not require any formal elaboration. What seems more problematic is why God did not carry out the punishment (death) He had previously threatened Adam and Eve with for eating the forbidden fruit.

The reason, I believe, lies in the fact that Adam, after a brief attempt at evasion by hiding in the garden of Eden (Gen. 3:8–13), admits his guilt. Moreover, both he and Eve could plead extenuating circumstances due to the misrepresentations of the serpent. Eve's claim to being duped seems to have been accepted, for the Bible does not report the serpent as entering a counterplea in its defense. In fact, the serpent is never heard from again.

In the Temptation Game (game 5 in the appendix), I have not included the serpent as a player; instead, I have made SB the player who chooses to tempt (T) or not tempt ($\overline{T}$) Eve. The serpent, like Satan in Job, seems no more than an agent of evil designed to test human subjects. P can resist pleasure (R) or not resist it ($\overline{R}$). I assume the primary and secondary goals of the players are as follows:

---

5. From another passage in the Bible (Ezek. 28:11–19), however, it appears that God was less upset by the moral knowledge that Adam and Eve gained from eating the forbidden fruit than by the hubris they displayed in daring to defy Him.

SB: (i) Wants P to resist pleasure; (ii) prefers to tempt P.

P: (i) Wants not to resist pleasure; (ii) prefers to be tempted.

Clearly, the primary goals of the players are diametrically opposed, with P wanting to resist pleasure and SB not wanting her to do so. Echoing the Bible, the tempter in different versions of the Faust legend is a devil (sometimes disguised), who promises youth, wealth, or some other undeserved indulgence.

If the primary goals of the players in the Temptation Game clash, their secondary goals do not: SB prefers to tempt P, and P prefers to be tempted. Unlike the Constraint Game, the conflict in the Temptation Game is only partial, with one player doing best when the other player does next worst ((4,2) and (2,4)), or one player doing next best when the other does worst ((3,1) or (1,3)). This game is depicted in figure 6.2.

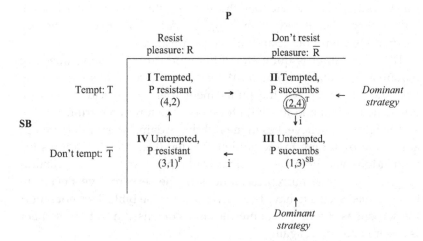

Key: (x,y) = (payoff to SB, payoff to P)
    4 = best; 3 = next-best; 2 = next-worst; 1 = worst
    i = impediment
    Nash equilibrium underscored
    Nonmyopic equilibrium (NME) circled
    Arrows indicate progression of states in weakly cyclic game with two impediments
    Superscripts P and SB indicate outcomes each player can induce with moving
        power; T is the threat outcome P can induce with a compellent threat

Figure 6.2
Outcome and payoff matrix of the Temptation Game (game 5)

SB's dominant strategy is T, and P's dominant strategy is $\overline{R}$, yielding the outcome (2,4), which is best for P but only next worst for SB. This is the unique Nash equilibrium in the Temptation Game. It occurs in virtually all versions of the Faust legend, though the person who succumbs to temptation enjoys only fleeting, and sometimes no, pleasure.

This outcome is also the unique NME, so wherever play starts, the players will be attracted to it. Even if SB does not tempt P and P is resistant at (3,1), P will move to (1,3), at which point SB will move to (2,4), where stoppage will occur, although SB would have preferred to stay at the initial state of (3,1).[6]

Can moving power help SB? First, observe that the Temptation Game is weakly cyclic, with two impediments, in a clockwise direction. Thus, compared with Belief Game 2 (see section 3.2) and the Constraint Game (see section 5.2), which are moderately cyclic, the players in the Temptation Game have less incentive to cycle indefinitely because of the greater friction caused by the two impediments. Nonetheless, if cycling occurs and SB has moving power, he can force P to stop at either (4,2) or (1,3), when P has the next move. Because P would prefer (1,3), this is the outcome she would choose, which, paradoxically, is worst for SB.

If P has moving power, she can force SB to stop at either (2,4) or (3,1), when he has the next move. Because SB would prefer (3,1), this is the outcome he would choose, which, also paradoxically, is worst for P. As in the Constraint Game, therefore, moving power is ineffective—it hurts the player who exercises it. Indeed, it hurts both players, because the moving power outcomes, (1,3) for SB and (3,1) for P, are Pareto-inferior to (2,4) and (4,2), respectively, which was not true of the moving power outcomes, (3,2) and (2,3), in the Constraint Game.

Besides moving power and order power, whose effect we analyzed in Belief Game 2 when the initial state is (1,2) (see section 4.3), there is a third kind of power—*threat power* (Brams 1994, chap. 5; 2011, chap. 7)—that P can exercise in the Temptation Game. It is of the *compellent* variety and involves P committing to her strategy of $\overline{R}$, compelling SB to choose between the Pareto-inferior state, (1,3), and the Pareto-superior state, (2,4). Because SB prefers (2,4) to (1,3), it is in SB's interest to choose T to make (2,4) the outcome.

---

6. Actually, SB would prefer to move to (4,2) from (3,1); the problem is that if he does so, P will countermove to (2,4), where, according to TOM, play will terminate.

If one player (the threatener) has threat power, he or she can force the other player (the threatened player) to choose between a Pareto-superior outcome and a Pareto-inferior outcome (*breakdown outcome*). If the threatened player does not choose a strategy that implements the Pareto-superior outcome, the threatener, in order to be credible, must carry out the threat.

Then both players, not just the threatened player, suffer at the breakdown outcome. A threat would not be necessary if only the threatened player suffered, because then the threatener could simply choose the action that leads to his or her preferred outcome. The point of a threat is that both players suffer if it is carried out, whereas both benefit if this does not happen.

P's compellent threat in the Temptation Game enables P, by choosing $\overline{R}$ and refusing to budge from it, to force SB to choose between the players' mutually preferred (2,4) state and their mutually nonpreferred (1,3) state. Note, however, that (2,4) is only SB's next-worst state. Might he, therefore, try to defy P?

He might indeed, but observe that SB cannot respond by compelling P to choose between a Pareto-superior and a Pareto-inferior state. If SB chooses T, perhaps hoping to implement his best state, (4,2), P will choose $\overline{R}$, inducing SB's next-worst state, (2,4); if SB chooses $\overline{T}$, P will again choose $\overline{R}$, inducing SB's worst state, (1,3). In other words, even if SB has a greater ability to withstand a Pareto-inferior state, he cannot, by choosing one of his strategies, induce P to choose a strategy that is in his (i.e., SB's) interest, much less in the players' mutual interest.

Thus, even though both players have dominant strategies in the Temptation Game, the structure of the game favors P at the unique Nash and nonmyopic equilibrium. While the possession of moving power helps neither player—inducing their worst outcomes at (3,1) or (1,3)—P benefits at (2,4) if she has compellent threat power, whereas SB cannot implement his best outcome at (4,2) with threat power.

But what does it mean for P to have threat power? Can P really threaten SB? It seems more accurate to say that, because SB has a dominant strategy of tempting P, P has even more reason to choose her own dominant strategy of not resisting such temptation. In fact, whatever the initial state, the players will, according to TOM, be drawn to (2,4) because it is the unique NME.

Ostensibly, this is not a good outcome for SB. Perhaps, however, SB views this outcome as not so bad, because it provides him with a justification for punishing P. Punishment sets an example for others in the future—namely, that succumbing to pleasure in the Temptation Game has consequences, even though it is an immediate setback for SB. In the longer run, however, SB will be happy if, in later games, P takes this as a precedent and, consequently, is more likely to be obedient.[7]

6.4 Conclusions

The failure of constraints in the Constraint Game makes it difficult to decide whether SB is a player, capable of getting his way when conflict is total. To be sure, in the Bible, God did punish Adam and Eve for their defiance of His authority. But punishment after the fact is not as strong evidence of God's powers as would be His prevention of a sin before it happens.

Consequently, the Constraint Game raises the question of whether SB is (i) a force to be reckoned with, (ii) an ordinary player like P, or perhaps (iii) not a player at all, especially because his exercise of moving power hurts rather than helps. Similarly, in the Temptation Game, SB cannot use his moving power to induce P to resist if tempted.

P's dominant strategy of not resisting pleasure in the Temptation Game can induce her best outcome if her compellent threat, which SB does not possess, of committing to this strategy convinces SB to back off. Moreover, wherever play starts, it is rational for the players to move and countermove to reach the Nash and nonmyopic equilibrium of (2,4), which gives SB only his next-worst outcome.

If this outcome has a saving grace, it provides SB with the opportunity to teach P the folly of disobedience.[8] It also affords SB, by punishing her, the opportunity to emerge as a stern but merciful disciplinarian, setting an example for future generations that would not have been possible had P not been tempted and had she adhered to the constraints that SB imposed.

---

7. SB's preferences in the Temptation Game do not reflect the longer-term consequences of his setting a precedent by punishing the offender. I explore these consequences more fully in games analyzed in later chapters, wherein SB gains considerable satisfaction from teaching his subjects a lesson.

8. But the learning may not be quick. It took twelve plagues before Pharaoh let the Israelites escape their servitude in Egypt (see section 9.3), after which he changed his mind, but with fatal consequences for himself and the pursuing Egyptians.

All in all, P's defiance of SB—by not adhering to constraints in the Constraint Game and by not resisting pleasure in the Temptation Game—disappoints SB in the short run. Indeed, these choices appear to undermine his authority, calling into question his decidability. However, they may have favorable long-term benefits by sending a signal that P's unrestrained exercise of her free will may have untoward consequences.

7   Three Testing Games

7.1   Introduction

God relentlessly tests the faith of His subjects in the Bible, as we saw with Adam and Eve in chapter 6. Even though they disobeyed God's command not to eat the forbidden fruit under threat of death, they were not killed for their transgression. Similarly, Job, a righteous man who did not violate any constraint imposed by God, nevertheless was horribly punished, though eventually he was richly rewarded for upholding his faith. I will analyze Job's complicated feelings and choices later in this chapter, but first I turn to two stories involving the possible sacrifice of children by a parent, perhaps the ultimate test of one's faith in God.[1]

Testing Game 1 is modeled after God's command to Abraham that he prepare his son, Isaac, for sacrifice. Fortunately for Abraham, an angel intervenes to prevent the sacrifice from being carried out. Testing Game 2 is modeled after a warrior, Jephthah, who makes a vow that he will, as an offering to God, kill the first living creature he sees if God helps him win a battle against the Ammonites. Unfortunately for Jephthah, after winning the battle, his daughter is the first person he lays eyes on.

Both games contain two "cooperative outcomes," (3,4) and (4,3), the first associated with SB's dominant strategy, the second with P's dominant strategy. Structurally, the games are the same: Interchanging the players transforms one game into the other, causing them to cycle in different directions but not changing their strategic features.

---

1. A novel, *Sophie's Choice* (Styron 1979), which was made into an award-winning movie in 1982, offers a searing portrait of a woman who must make a choice during the Holocaust between which of her two children will be killed.

While one of the (3,4) and (4,3) outcomes in each game is a Nash equilibrium, both of these outcomes are NMEs. This puts these outcomes in a kind of tug of war, in which the initial state, as well as different kinds of power, can make a difference as to which will be chosen by the players. The outcome in Testing Game 1 (Abraham's attempted sacrifice) depends on the prediction SB makes about P's choice—and P's awareness of SB's prediction—whereas SB's prediction in Testing Game 2 (Jephthah's sacrifice) has no effect on P's choice.

Besides Testing Games 1 and 2, I analyze a third testing game, Testing Game 3, which is identical to Testing Game 2 though its strategies have a very different interpretation. It models Job's precarious situation when God, after some hesitation, permits Satan to test his faith. Satan proceeds, first, by annihilating Job's family and then subjecting Job to terrible deprivation. While expressing great anguish during his grueling test, Job passes it and is ultimately rewarded for not losing his faith in God.

I suggest that as Job becomes more desperate, he must make a calculation about whether God will come to his rescue, in the end, by reining in Satan. Reinterpreting this game in terms of choices by P and SB, it turns out in Testing Game 3 to be irrelevant what SB predicts P will do—if P is tested, she should do everything in her power to pass the test, despite the enormous pain she may suffer.

In the Bible story represented by Testing Game 1, Abraham seems to have anticipated that God's command to sacrifice Isaac was only a test. Consequently, he could go through the motions of preparing Isaac for sacrifice without fear that the sacrifice would occur.

On the other hand, in Testing Game 2, God did not view as revocable Jephthah's vow to sacrifice the first living creature he sees. Hence, God made no attempt to arrest the sacrifice of Jephthah's daughter, though He allowed her sacrifice to be postponed. God also made no attempt to stop Satan from wiping out Job's entire family in Testing Game 3, but unlike Jephthah, God subsequently blessed Job with a new family and vast riches.

The fact that biblical characters are tested, sometimes to their wit's end, may strike one as not only overkill but also capricious. But recall that if we indeed have free will, there is nothing to constrain us from doing untoward things, at least in the eyes of God. Tests of faith help to ensure that loyalty is ultimately rewarded and disloyalty, or a lack of zeal in expressing one's faith and upholding one's vows, is punished. Furthermore, testing under extreme duress sets a precedent from which others may learn not to try to cross God.

7.2   Testing Game 1: Abraham's Attempted Sacrifice of Isaac[2]

With characteristic economy of language, chapter 22 of Genesis begins, "Some time afterward, God put Abraham to the test" (Gen. 22:1). Then, in just eighteen verses, one of the greatest and most poignant stories in the Bible is told. The significance of this story, and its interlocking themes of faithfulness, sacrifice, and possible murder, have been subject to prodigious analysis and interpretation, some of which I will briefly discuss later.

In the story, God commands Abraham:

Take your son, your favored one, Isaac, whom you love, and go to the land Moriah, and offer him there as a burnt offering on the heights which I will point out to you. (Gen. 22:2)

Faithful servant of God that Abraham is, he sets out on his ass with Isaac, accompanied by two of his men with firewood for the sacrifice.

On the third day of the journey, Abraham sees the place for the sacrifice and leaves his ass and two men behind. He gives Isaac the firewood to carry, and he himself carries the firestone and the knife. When Isaac asks, "Here are the firestone and the wood, but where is the sheep for the burnt offering?" (Gen. 22:7),[3] Abraham answers, "God will see to the sheep for His burnt offering my son" (Gen. 22:8).

Abraham builds an altar and lays out the wood, after which he binds Isaac and lays him on the altar on top of the wood. As he picks up his knife to kill his son,

An angel of the LORD called to him from heaven: "Abraham! Abraham!" And he answered, "Here I am." And he said, "Do not raise your hand against the boy, or do anything to him. For now I know that you fear God, since you have not withheld your son, your favored one, from Me." When Abraham looked up, his eye fell upon a ram, caught in the thicket by its horns. So Abraham went and took the ram and offered it up as a burnt offering in place of his son. (Gen. 22:11–13)

Abraham is then rewarded for his faithfulness—or at least his obedience to authority—when the angel calls from heaven a second time:

---

2. This section and the next are based in part on Brams ([1980] 2003, chap. 3).

3. Isaac's question is the title, *But Where Is the Lamb? Imagining the Story of Abraham and Isaac*, of Goodman's (2013) inquiry into how this question has been understood and analyzed by scholars and commentators from different religious and philosophical traditions. Pawlowitsch (2012) proposed an extensive-form game of incomplete information to explain the choices of Abraham and God, wherein God signals to Abraham the equilibrium he wants played.

By myself I swear, the LORD declares: because you have done this and have not withheld your son, your favored one, I will bestow My blessing upon you and make your descendants as numerous as the stars of heaven and the sand on the seashore; and your descendants shall seize the gates of their foes. All the nations on earth shall bless themselves by your descendants, because you have obeyed my command. (Gen. 22:16–18)

I base Testing Game 1 on Abraham's attempted sacrifice, but as I have done before, I make SB and P the players—instead of God and Abraham—to model a testing game that mimics the choices these biblical characters faced. After SB commands P to make a human offering, which could be any other life-or-death request, P can offer (O) or not offer ($\overline{O}$) to make such an offering; in turn, SB can renege (R) or not renege ($\overline{R}$) on his command. I assume the primary and secondary goals of the players are the following:

SB: (i) wants to renege on his command (because this is only a test); (ii) prefers that P obey his command and make a human offering (to show her faith).

P: (i) wants SB to renege on his command; (ii) prefers not to obey SB's command and make a human offering.

The resulting game, which is game 33 in the appendix, is shown in figure 7.1.[4] Based on her primary goal, P's two best outcomes (4 and 3) are associated with R and her two worst outcomes (2 and 1) with $\overline{R}$. Secondarily, P would prefer not to make a human offering, so 4 and 2 are associated with $\overline{O}$ and 3 and 1 are associated with O. SB also prefers R (3 and 4) to $\overline{R}$ (1 and 2), but unlike P, he prefers that P make an offering, so 4 and 2 are associated with O, and 3 and 1 are associated with $\overline{O}$.

Thus, there is agreement that SB should choose R (both players' two best outcomes are associated with this strategy), but there is a clash over whether P should obey SB, which SB prefers, or not obey, which P prefers. As figure 7.1 shows, there are two competing Pareto-optimal outcomes—(4,3) favors P, (3,4) favors SB. Game theory predicts (4,3), because it is the product of P and SB's strongly dominant strategies and is the unique Nash equilibrium in this game.

By contrast, from (3,4) P would benefit by switching from O to $\overline{O}$, which yields (4,3). Although SB cannot improve on his next-best

---

4. This game is one of four "king-of-the-mountain games," which are related to "catch-22 games"; both classes of games are analyzed and compared in Brams and Jones 1999 and Brams 2011, chap. 10.

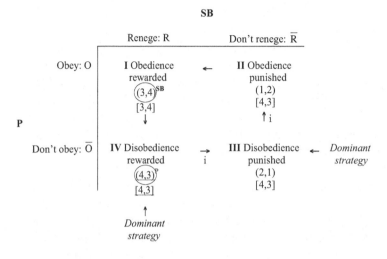

Key: (x,y) = (payoff to SB, payoff to P)
    4 = best; 3 = next-best; 2 = next-worst; 1 = worst
    i = impediment
    Anticipation game given in brackets
    Nash equilibria underscored
    Nonmyopic equilibria (NMEs) circled
    Arrows indicate progression of states in weakly cyclic game
    Superscripts P and SB indicate outcomes each player can induce with moving
        power or threat power

Figure 7.1
Outcome and payoff matrix of Testing Game 1 (game 33)

outcome of 3 by switching to $\overline{R}$, the fact that Abraham can do better prevents (3,4) from being a Nash equilibrium, much less a dominant-strategy Nash equilibrium (note that O is a strongly dominated strategy for P).

Both (4,3) and (3,4) are NMEs, from which neither player would move (unlike boomerang NMEs; see section 4.3); play will migrate to the former from (2,1) and to the latter from (1,2). If either player has moving power, he or she can implement his or her preferred NME, making the possession of moving power effective. Each player can also implement his or her preferred NME with threat power, which I will discuss shortly in this game.

In the actual Bible story, the Nash equilibrium, in which Abraham does not offer Isaac but God reneges anyway, was not chosen. Abraham did offer Isaac, and only later did God renege on His command and prevent Isaac's sacrifice.

Several modern commentators consider Abraham's decision, despite its favorable consequences, odious. They believe that Abraham should have pleaded for Isaac's life, as he did for saving the inhabitants of Sodom and Gomorrah. Other commentators consider Abraham's attempt to sacrifice Isaac a morally reprehensible act.[5] If Abraham's preferences mirror P's in Testing Game 1, O is a dominated strategy, so, from the perspective of standard game theory, Abraham should *not* have obeyed God's command to sacrifice Isaac.

How can one reconcile Abraham's concern for his son's life, which I postulated earlier in assuming that he preferred $\overline{O}$ to O, with his choice of offering Isaac? First, recall Abraham's answer to Isaac's plaintive question, "But where is the sheep for the burnt offering?" (Gen. 22:7): "God will see to the sheep for His burnt offering, my son" (Gen. 22:8). Abraham's response strongly suggests that he believed Isaac would not be sacrificed and, moreover, did not want to alarm Isaac or later the servants.

As further evidence of Abraham's belief that he was only being tested, Abraham had told his servants, before leaving them and the ass behind, that "we will return to you" (Gen. 22.5). This apparent foreknowledge of Abraham that God would renege on his command that Isaac be sacrificed seems one reason why Abraham was reluctant to defy God's edict or plead for Isaac's life.

But there is another cogent reason why Abraham was willing to choose O. On several previous occasions, God had made extravagant promises to Abraham, telling him, among other things,

I will make of you a great nation,
And I will bless you;
I will make your name great;
And you shall be a blessing. (Gen. 12:2)

I will make your offspring as the dust of the earth, so that if one can count the dust of the earth, then your offspring too can be counted. (Gen. 13:16)

"Look toward heaven and count the stars, if you are able to count them." And He added: "So shall your offspring be." (Gen. 15:5)

I will make you the father of a multitude of nations. I will make you exceedingly fertile, and make nations of you; and kings shall come forth from you. I will maintain My covenant between Me and you, and your offspring to come,

---

5. An excellent summary of these views can be found in Dershowitz 2000.

as an everlasting covenant through the ages, to be God to you and to your offspring to come. (Gen. 17:5–7)

In speaking of Abraham's wife, Sarai (later called Sarah), who had been barren for many years, God said to Abraham:

I will bless her; indeed, I will give you a son by her, I will bless her so that she shall give rise to nations; rulers of people shall issue from her. (Gen. 17:16)

That son, of course, was Isaac, whom Sarah bore at the age of ninety (Abraham was then one hundred). God added,

I will maintain My covenant with him [Abraham] as an everlasting covenant for his offspring to come. (Gen. 17:19)

Given all these assurances, is it conceivable that Abraham believed that God meant for him to sacrifice Isaac, the progenitor-to-be of multitudinous offspring?

My answer is that it is conceivable, given the steadfastly God-fearing man that Abraham appears to have been. (Alternatively, he lacked the moral fortitude to stand up to God to try to save Isaac's life, as some of the commentators cited earlier suggest.) But it is also conceivable that Abraham suspected that he was only being tested and calculated that if this was the case, it was rational for him to offer Isaac.

Abraham was not above this kind of thinking, as illustrated by a story from his past in which he tried to pass off Sarah as his sister. Because of Sarah's great beauty, Abraham feared that if it were known that he was her husband, he would be killed by the Egyptians, and the Pharaoh at the time could take Sarah as his wife.[6] In fact, Pharaoh did take Sarah as his wife when Abraham lied that she was his sister, but when the truth came out, Pharaoh angrily ordered the dissembling couple out of Egypt.

I offer all this as background to make the point that while it is hard to say exactly what game Abraham was playing, it is certainly possible to imagine that Abraham surmised that God preferred that he offer— but not actually sacrifice—Isaac. While God's harrowing test of Abraham succeeds in establishing that Abraham would obey His command—however ghastly—Abraham may well have done so for

---

6. For an alternative motive that emphasizes Abraham's desire to protect his wife rather than save himself, see *Anchor Bible: Genesis* (1964, xl–xli). Whatever Abraham's precise motive may have been, he appears not to have been averse to trying to deceive Pharaoh.

reasons other than faith. Hence, God's test does not dispel doubts about Abraham's faith, given that Abraham knows God's preferences and is rational.[7]

Aside from Abraham's story, is there anything in Testing Game 1 that would suggest that P not choose his strongly dominant strategy of $\overline{O}$ and instead choose O? If SB has *deterrent threat power*, he can threaten a Pareto-inferior outcome—called a *breakdown outcome* (see section 6.3)—that is worse for both players than a Pareto-optimal outcome. Not only can a player with threat power threaten such a breakdown outcome, but also this player can, I assume, hold out longer at it than can a player without such power.

Assume that SB has threat power in Testing Game 1. Then he can threaten to choose $\overline{R}$, which would lead either to P's worst outcome at (1,2) or her next-worst outcome at (2,1). Facing this choice, P presumably would choose $\overline{O}$ to effect (2,1). But this outcome is worse for P, as well as SB, than (3,4), which is a Pareto-optimal *threat outcome* that both players prefer to breakdown outcome (2,1). In the face of a threat that SB might choose $\overline{R}$ if P chooses $\overline{O}$, yielding (2,1), it is rational for P to choose O to induce (3,4).[8]

---

7. Alternatively, Abraham's faith may have been fueled by his *fear* of God as much as his faith, which is most prominently discussed in Kierkegaard's *Fear and Trembling* ([1843] 1954). But the Bible provides insufficient information to say whether Abraham's possible blind faith was fear-induced. The element of fear, and a compliant Abraham, is expressed in the lyrics of Bob Dylan's song, "Highway 61 Revisited":

Oh God said to Abraham, "Kill me a son"

Abe says, "Man, you must be puttin' me on"

God say, "You can do what you want, Abe, but

The next time you see me comin' you better run,"

Well, Abe says, "Where do you want this killin' done?"

© 1965 Warner Bros. Inc. All rights reserved. Used by permission.

By comparison, here is how Woody Allen (2007, 138) injects black humor into the dialogue between Abraham and God:

At the last minute the Lord stayed Abraham's hand and said, "How could thou doest such a thing?"

And Abraham said, "But thou said—"

"Never mind what I said," the Lord spake. "Doth thou listen to every crazy idea that comes thy way?" And Abraham grew ashamed. "Er—not really ... no."

"I jokingly suggest that thou sacrifice Isaac and thou immediately runs out to do it."

And Abraham fell to his knees, "See, I never know when you're kidding."

And the Lord thundered, "No sense of humor. I can't believe it."

8. If P were in a position to threaten SB and had threat power, she has a compellent threat to choose $\overline{O}$, which would induce (4,3), her preferred NME (see section 6.3), as compared with SB's deterrent threat that induces (3,4). But there is no evidence that

be wracked by guilt that he might have saved his son by acting differently.

For these reasons, I believe, Abraham probably could have gotten away with refusing to sacrifice Isaac. After all, God's threat was never explicit, so there would not have been an enormous need for face-saving on God's part. Furthermore, God evinced later a willingness to bargain with Abraham, listening to Abraham's appeals on behalf of the cities of Sodom and Gomorrah and offering them reprieves, even though, in the end, neither had enough righteous inhabitants to be worth saving.

I believe God would have been more open to an appeal on behalf of an innocent child than the wicked inhabitants of Sodom and Gomorrah, even if Abraham's refusal to offer Isaac was not what God most wished to happen. From a moral standpoint, Abraham's refusal would have shown the he had the courage to stand up for something of paramount importance to himself, just as Moses, while infuriated by the behavior of the Israelites for building the golden calf (see section 9.4), still stood up for their survival (though some Israelites were slaughtered for their idolatry). But, I conclude, Abraham was no Moses, about whom it was said, "Never again did there arise a prophet like Moses" (Deut. 34:10).

Testing Game 1, and the preferences on which it is built, suggests that Abraham's refusal would *not* have been catastrophic either for him or for God. Contrary to what happened in one of the most wrenching situations to face a character in the Hebrew Bible, I believe the counterfactual can be entertained. More specifically, if Abraham was indeed the caring father, he (rationally) could have ignored God's implicit threat, or at least pleaded for Isaac's life, which doubtless would have solidified his reputation as not only the patriarch of the Jews but also a morally upright one.

7.3 Testing Game 2: Jephthah's Sacrifice of His Daughter

As told in chapter 11 of Judges, Jephthah, an intrepid warrior and the son of a prostitute, was driven from his home in Gilead by the legitimate sons of his father. In the country of Tob where he settled, "men of low character gathered about Jephthah and went out raiding with him" (Judg. 11:3). This problem notwithstanding, the elders of Gilead, faced by an Ammonite attack, recalled him and sought his aid, which he consented to give on the condition that they appoint him commander of Gilead after his victory.

The elders agreed, and Jephthah then tried to negotiate with the Ammonites, but the negotiations broke down. Forced into battle, Jephthah made the following fateful vow to the LORD:

If you deliver the Ammonites into my hands, then whatever comes out of the door of my house to meet me on my safe return from the Ammonites shall be the LORD's and shall be offered by me as a burnt offering. (Judg. 11:30–31)

Having made this vow, Jephthah routed the Ammonites, who then submitted to the Israelites. Upon returning to his home, Jephthah, to his utter dismay, was greeted by his daughter and only child "with timbrel and dance" (Judg. 11:34). His heart broken, Jephthah rent his clothes and told his daughter, "I have uttered a vow to the LORD and I cannot retract" (Judg. 11:35).

Resigned to her fate, Jephthah's daughter dolefully asked that her sacrifice be postponed for two months so that, with her companions, she could "lament upon the hills and there bewail my maidenhood" (Judg. 11:37). With God's apparent consent, Jephthah granted her this wish, but at the end of this period he grimly fulfilled his vow. The Bible reports that it henceforth became a custom in Israel for maidens to commemorate this tragic event by chanting dirges for the daughter of Jephthah during an annual four-day mourning period.

I base Testing Game 2 on Jephthah's sacrifice, but as with Abraham's sacrifice, I make SB and P the players instead of God and Jephthah. After P makes a vow to offer (O) or not offer ($\overline{O}$) to make a burnt offering if he is successful in battle—or some other endeavor—SB may either stop (S) or not stop ($\overline{S}$) P from upholding her vow. I assume that goals of the players are the following:

SB: (i) wants P to uphold her vow to offer; (ii) prefers not to stop P from carrying out her vow.

P: (i) wants to uphold her vow to offer; (ii) prefers that SB stop her from carrying out her vow.

The resulting game is shown in figure 7.2. As with Abraham's attempted sacrifice, the players agree on their primary goal, in which each player's two best outcomes are associated with O. But they disagree on their secondary goals: Whereas SB does not want to stop the sacrifice, P desires that he do so.

In fact, this is the same game (game 33) as that shown in figure 7.1, but with the roles of the players reversed. Now (4,3) and (3,4) are associated with one of P's strategies (O), whereas the two best outcomes in

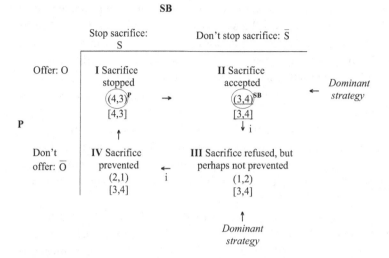

Key: (x,y) = (payoff to SB, payoff to P)
 4 = best; 3 = next-best; 2 = next-worst; 1 = worst
 i = impediment
 Anticipation game given in brackets
 Nash equilibria underscored
 Nonmyopic equilibria (NMEs) circled
 Arrows indicate progression of states in weakly cyclic game
 Superscripts P and SB indicate outcomes each player can induce with moving
    power or threat power

Figure 7.2
Outcome and payoff matrix of Testing Game 2 (game 33)

Testing Game 1 are associated with one of SB's strategies (R). Similarly, the players' two worst outcomes, (2,1) and (1,2), are associated with SB's $\overline{R}$ in Testing Game 1 but with P's $\overline{O}$ in Testing Game 2.

Because Testing Game 1 and Testing Game 2 are structurally the same, our analysis of Testing Game 1 is applicable to Testing Game 2, except for the interchange of players. However, the outcome for the biblical players in Testing Game 1—Abraham offers Isaac, and God reneges on this command—does not correspond to the outcome that occurs in Testing Game 2 when Jephthah upholds his vow to sacrifice his daughter, and God does not stop him.[12]

---

12. If P does not offer to make a sacrifice in Testing Game 2, there would be no sacrifice to stop, rendering SB's strategies of stopping or not stopping the sacrifice meaningless. But it seems that Jephthah had no recourse other than to offer to make a sacrifice if he wanted to win the battle against the Ammonites—making his choice to do so a *feasible move* (that is, one that is possible)—which is consistent with O's being a dominant strategy.

This reversal of outcomes in the two Bible stories requires explanation. Each testing game has the same two NMEs, (3,4) and (4,3), but in Testing Game 1, the NME that is not a Nash equilibrium was chosen in the Abraham story, whereas the NME that is a Nash equilibrium was chosen in the Jephthah story. To understand why in the Jephthah story, we need to look more closely at the situation that Jephthah faced when he made his vow.

Since Abraham's death, the Israelites had caused God much grief, both before and after the conquest of Canaan by the Israelites (Brams [1980] 2003, chaps. 5 and 7). Consequently, God was not inclined to be sympathetic with people, like Jephthah, who were quick to make solemn vows.

Whether Jephthah could anticipate this problem or only remembered the test of Abraham is hard to say. Because it is believed that Jephthah's story was written before Abraham's, however, it is questionable whether precedents are meaningful here.

But would Jephthah have acted differently if he foresaw that his daughter would be the first person to emerge from his house after his victory? It is hard to say, though clearly Jephthah was devastated when he realized the terrible consequences of his vow. Nevertheless, he still carried out the sacrifice.

In contrast, Abraham seemed not at all perturbed about sacrificing his son. But as I argued earlier, Abraham seems to have foreseen that he was only being tested and so would not have to consummate the sacrifice.

In summary, the players chose two different NMEs in Testing Games 1 and 2. God was merciful in Abraham's case, but he was more vindictive in the case of Jephthah. To be sure, Jephthah's vow made him more culpable than Abraham. However, without God's blessing, Jephthah probably would have failed to defeat the Ammonites,

Anticipating that he would not survive the battle, Jephthah had good reason to make the vow he did. At least God was willing to tolerate some delay before Jephthah had to dispatch his daughter, though that seems to have given little comfort to either Jephthah or his daughter.

The different outcomes in Testing Games 1 and 2, which are structurally the same despite the interchange of players and their preferences, are both NMEs and so are predicted by TOM. But the fact that each can occur may leave unclear who SB is and what he wishes to accomplish. The final testing game does not do much to clear up this mystery.

7.4   Testing Game 3: Job's Suffering at the Hands of Satan

The Book of Job begins by presenting Job as a righteous and virtuous man:

There was a man in the land of Uz named Job. That man was blameless and upright; he feared God and shunned evil. (Job 1:1)

Although Job was the wealthiest man in Uz, he still would, "rising early in the morning, … make burnt offerings" (Job 1:5), including ones for each of his children.

But trouble brews when Satan asks God:

Does Job not have good reason to fear God? Why, it is You who have fenced him round, him and his household and all that he has. You have blessed his efforts so that his possessions spread out in the land. But lay Your hand upon all that he has and he will surely blaspheme You to Your face. (Job 1:10–11)

Even after all Job's children are killed by a "mighty wind"—seemingly instigated by Satan—that strikes and collapses his eldest son's house, Job keeps his faith, saying that "the Lord has given and the Lord has taken away; blessed be the name of the Lord" (Job 1:21).

When Satan approaches God a second time, God, as He did earlier, extols Job's faith and says that Job has kept his integrity, despite the fact that "you [Satan] have incited Me against him to destroy him for no good reason" (Job 2:3). Nevertheless, when Satan again challenges Job's faith, this time God allows Satan to put Job in his power, though Satan must spare Job's life (earlier God had forbidden Satan to "lay a hand on him" but not his family) (Job 1:12).

The Bible recounts in great detail all the depredations Satan inflicts on Job, including great physical pain and mental suffering. Three friends try to comfort Job, but to no avail. On several occasions Job falters in his faith, saying such things as:

On my part, I will not speak with restraint;
I will give voice to the anguish of my spirit;
I will complain in the bitterness of my soul. (Job 7:11)

Job even questions God's role in sanctioning his suffering, saying to his friends,

Yet know that God has wronged me;
He has thrown up siege works around me.
I cry, "Violence!" but am not answered;
I shout, but can get no justice. (Job 18:6–7)

Job beseeches his friends, "Pity me, pity me!" (Job 19:21). But he also says, "I persist in my righteousness and will not yield" (Job 27:6).

Job's steadfastness pays off when, eventually, Satan's attempt to provoke him into abandoning God fails. Job is rewarded not only with great wealth but also a new family (seven sons, and three daughters whose beauty is surpassed by no other women in the land). He "lived one hundred and forty years to see four generations of sons and grand-sons" and died "old and contented" (Job 42:16–17).

Like Abraham, Job passed a gruesome test of his faith in God. But if Abraham was anguished, he did not express it, whereas Job cried out in pain and suffering. Whether Abraham's stoicism or Job's pleas for relief—both while keeping their faith—is more praiseworthy is difficult to say, though some would claim that nobody deserves to be tested in the way Job was.

I assume in Testing Game 3 (see figure 7.3) that SB can either test (T) or not test ($\overline{T}$) P. If tested, P can choose either to endure (E) or not endure ($\overline{E}$) the test. If not actually tested, P can still decide whether she would try to endure a test.

I assume the goals of SB and P in Testing Game 3 are the following:

SB: (i) wants P to endure any test by Satan; (ii) prefers that P be tested.

P: (i) wants to endure any test by Satan (ii); prefers not to be tested.

In fact, this game is identical to Testing Game 2, except for the inter-change of SB's two strategies in Testing Game 3, which changes the direction of cycling from clockwise in Testing Game 2 to counterclock-wise in Testing Game 3. As in Testing Game 2, SB's powers of predic-tion, and P's awareness of these, have no effect on P's choice of her dominant strategy.[13]

As in Testing Game 2, the NME in Testing Game 3 that is a Nash equilibrium, (3,4), is the outcome that was chosen by Job and Satan (as SB) in the Bible. It is also the outcome that SB can induce with moving power, which it is reasonable to suppose Satan had.

The game starts at (4,3), before Job has been tested, and his integrity is intact. As the testing subjects Job to excruciating pain and depriva-tion, he contemplates choosing $\overline{E}$. It seems probable that God contem-

---

13. A player's choice of her dominant strategy may not always lead to her preferred outcome, as I showed for Newcomb's problem in section 5.2.

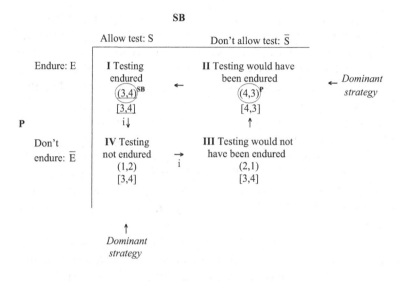

Key: (x,y) = (payoff to SB, payoff to P)
    4 = best; 3 = next-best; 2 = next-worst; 1 = worst
    i = impediment
    Anticipation game given in brackets
    Nash equilibria underscored
    Nonmyopic equilibria (NMEs) circled
    Arrows indicate progression of states in weakly cyclic game
    Superscripts P and SB indicate outcomes each player can induce with moving
        power or threat power

Figure 7.3
Outcome and payoff matrix of Testing Game 3 (game 33)

plated halting the test, but then Job shows his mettle, after faltering, by continually committing himself to God. The game cycles again and again through these states. If in Testing Game 3 SB has moving power, he can induce the choice of (3,4), wherein P endures all obstacles thrown at her during the testing.

One difference between Job's situation and Jephthah's is that God in the Jephthah story never tried to halt the sacrifice of Jephthah's daughter, presumably because He did not want to release Jephthah from the sacred vow he had voluntarily made. It would set a bad precedent if characters like Jephthah could receive God's help but, after doing so, not fulfill their part of a bargain they had struck.[14]

---

14. In a different bargaining situation, Abraham, to prevent the destruction of Sodom and Gomorrah, also had to accept terms he had agreed to when he could not find as few as ten righteous men in these cities.

Job, on the other hand, made no vow, nor did he attempt to bargain with God about his dreadful treatment by Satan, though he certainly did complain bitterly. Indeed, unlike bargaining with the devil in different versions of the Faust legend, there was no communication at all with Satan. Instead, the bargaining was between Satan and God, at the outset, over Job's fate.

In the case of Job, God was hesitant to permit Satan to test Job since Job had done nothing to deserve such harsh treatment. In the beginning, Satan was allowed to wreak havoc on Job's family but not touch Job himself; only after Job is unflinching when his children are killed—in what looks like a natural disaster (a mighty wind)—does God permit Satan to go after Job directly, which brings him to death's door and utter despair.

In the end, God's judgment about Job's faith is vindicated when Job withstands every indignity imaginable. But was it faith alone that sustained Job, or might he have predicted that all would be set right if he persisted in his loyalty to God?

Once the test began in Testing Game 3, I assume that P can surmise that she is at (3,4). Her only choice is to move to (1,2), which not only is worse for her but also for SB. Moreover, if SB subsequently moves to (2,1), he would suffer even more, so SB cannot do better by stopping the test.

Thus P, basing her choice not only on her own preferences but also on her knowledge of SB's preferences, does better by holding out at (3,4). Being tested is anything but ideal for P, but it is best for SB if, when the test has ended, P has succeeded. And we know that when Job's awful punishment did end, he was amply rewarded for the fortitude he showed in dire circumstances.

7.5 Conclusions

Human sacrifice, especially of family members, is no longer condoned in the civilized world, though it certainly occurs in some theocratic societies. In the Bible, God not only sanctioned such sacrifice, but sometimes He did nothing to prevent it from occurring.

His purpose in the case of Abraham and Job was to test the faith of two righteous individuals, and in each case they passed the test. One might question the morality of such a supreme test, which Abraham did not dispute but which Job definitely did. In doing so, Job bewailed his fate, whereas Abraham never uttered a word of disagreement with God's command.

Abraham's stoicism may be applauded or condemned, but I think the relevant strategic question is why he remained imperturbable. I suggested that he had good reason to believe that he was only being tested after God had offered him multiple assurances that he would found a great nation. Clearly, this nation could not survive if Isaac, the only child of Abraham and Sarah, were killed.

Some scholars, as I indicated earlier, have argued that Abraham should have spoken up, as Job did, about the immorality of God's command. Game theory makes no judgment about such a question but, instead, tries to explain why Abraham remained silent. The likely answer is that he probably anticipated that Isaac's sacrifice would be halted. By contrast, Job harbored no such sentiment, especially after suffering prolonged agony, which explains why he was so dispirited and almost, in a manner of speaking, threw in the towel: There seemed little prospect that his circumstances would improve.

Jephthah, too, despaired over his fate, but not because he was physically harmed. Rather, after he had made a solemn vow he could not retract, he was distraught about having to carry it out. But Jephthah's punishment served a larger strategic purpose: It signaled to others that they would be held accountable for thoughtless promises they made.

What light do the three stories of sacrifice shed on the undecidability of SB? In Testing Games 2 and 3, it does not matter if SB can predict, even with certainty, P's action, because there is no probability p that can induce P to choose, on the basis of an expected-value calculation, one strategy over the other (see section 5.4). The dominance of one of P's strategies in these games precludes such a choice.

In Testing Game 1, SB has a dominant strategy of reneging, but in the Bible, Abraham did not choose his own dominant strategy of disobeying God's command. Instead of showing his moral courage through disobedience, Abraham prepared to carry out God's command, in part, it seems, because he correctly surmised he was only being tested and would not, in the end, be forced to kill Isaac.

In Testing Game 1, a reason for P to obey is if she believes that SB's command carries an implicit threat to punish her if she does not obey, though it is difficult to estimate the severity of the punishment P would receive if she did not obey. For one thing, P's two best outcomes are associated with SB's dominant strategy, so this does not help P to decide which to choose, given her uncertainty about her punishment if she disobeys. Also, both the obey and disobey outcomes are NMEs and moving-power outcomes, so again each of P's choices seems viable.

In short, SB's superiority does not offer P an obvious avenue of escape from Abraham's having to make a fateful decision. Nevertheless, he managed it without possessing any supernatural powers but instead seems to have made a straightforward determination that it was not in God's interest to permit the sacrifice.

In Testing Game 2, both players again have dominant strategies, and the resulting outcome is both a Nash equilibrium and an NME, which SB can implement with moving power. Indeed, this is the outcome that occurred in the Bible when Jephthah sacrificed his daughter, but if P has moving power, she can induce SB to stop the sacrifice. Evidently, this was not true for Jephthah.

But the real uncertainty for Jephthah arose from not anticipating that the creature he would have to kill would be a member of his family. This disastrous circumstance has nothing to do with the fact that SB is superior; if Jephthah had made the same promise to a king, he still would have had to fulfill it, though there remains the question of whether a king would have been able to help Jephthah in his battle with the Ammonites.

In Testing Game 3, P has a dominant strategy of enduring any torture inflicted on her, and SB's best response—yielding the Nash equilibrium that is also an NME—is to test her (more accurately, allow her to be tested by Satan). SB can implement this outcome with moving power. P, of course, would prefer not to be tested; there is, in fact, a second NME associated with her not being tested but be willing to endure torture if she is. But God, pressured by Satan, let the test proceed.

As in Testing Games 1 and 2, the outcome chosen in the Bible in Testing Game 3 might well have been chosen if SB did not have superior powers. Both Pareto-optimal outcomes are NMEs, which is to say that no special power required to induce one. Hence, it is not apparent that SB's superiority is critical in any of the Testing Games, though some kind of power (moving or threat) might have reinforced the outcome that was chosen.

8   The Incitement, Blame, and Deception Games

8.1   Introduction

In the three Testing Games analyzed in chapter 7, SB's rulings are quite different: He issues a command and then later retracts it (Abraham's attempted sacrifice), he fails to stop a sacrifice (by Jephthah), and he allows a righteous person (Job) to be grievously tested, putting only minimal limits on the nature and severity of the test. The biblical God in each game does not express any strong emotions, except to indicate His pleasure that Abraham and Job pass their tests (God does not comment on Jephthah's sacrifice of his daughter).

Circumstances are not so benign for the three human characters. There is probably no more excruciating test of a person's faith than making his affirmation of God conditional on the sacrifice of a child. It is an unbearable choice for a parent, but when forced to choose, Abraham and Jephthah were ready to make that choice. As I showed in the case of Abraham, however, his affirmation may have been rooted in a rationality other than blind faith. Although Abraham never appears disturbed by God's command, the same cannot be said for Jephthah's heartbreak about having to fulfill his vow. Likewise, Job is devastated by the death of his first family and the later suffering he must endure.

I next turn to three biblical games in which God is not so impassive, as He mostly is in the three foregoing games, but is clearly angry with situations that, ironically, He creates. In the Incitement Game, He praises Abel's sacrificial offering but pays no heed to Cain's, which produces in Cain raging jealousy and incites him to murder his brother, Abel. In the Blame Game that immediately follows Abel's murder, Cain cleverly attempts to shift, at least partially, the blame for Abel's murder to God by suggesting that He could have prevented it. In the Deception Game, Saul, when he agrees reluctantly to become the first king of

Israel, is erroneously led to believe that he will be supported by the prophet, Samuel, and God. However, his rule is systematically undermined by these erstwhile supporters.

The undecidability problem in the games that are based on these stories stems in part from the fact that SB's superiority does not significantly change the outcomes that would occur if SB were an ordinary being, without any special powers. But SB's actions raise a different kind of issue, because they precipitate death or suffering in each story. One can, therefore, question the moral rectitude of SB's choices in fostering, rather than condemning, evil in the world.[1] If the morality of SB's choices is no better than P's, and maybe worse, then SB cannot be distinguished from P in terms of an ethical standard, further contributing to his undecidability.

8.2 The Incitement Game: Provoking Cain's Murder of Abel

After being driven from the garden of Eden, Adam and Eve become parents first to Cain and then to Abel. Cain grows up to be a tiller of the soil, Abel a shepherd. As if incapable of letting things take their own course after the expulsion of Adam and Eve from the garden of Eden, God set up the conditions for conflict between the two brothers:

In the course of time, Cain brought an offering to the LORD from the fruit of the soil; and Abel, for his part, brought the choicest of the firstlings of his flock. The LORD paid heed to Abel and his offering, but to Cain and his offering He paid no heed. Cain was much distressed and his face fell. (Gen. 4:3–5)

Unlike the Constraint Game discussed in section 6.2, this time God does not place limits on human choice and wait for them to be violated: He meddles directly in the affairs of the brothers by playing favorites, naturally antagonizing the one not favored.

True, Cain's offering was apparently inferior to Abel's, because it was simply from the "fruit of the soil" (Gen. 4:3) but not, like Abel's, the "choicest" (Gen. 4:4). But if God was disappointed by the meagerness of Cain's offering, why did He not say so, instead of paying no heed to Cain? God had not been silent about His distress with Adam and Eve's transgressions.

---

1. This issue is present but less apparent in the three Testing Games in chapter 7, because it can be claimed in each case that a test was required to distinguish the faithful from the unfaithful.

In fact, God's primary motive seems to have been less to chastise Cain than to alleviate His boredom by stirring up jealousy between the brothers—and then await the fireworks. If this was His goal, He was not to be disappointed.

As support for this position, consider God's incredible question after refusing Cain's offering and observing his anger:

Why are you distressed,
And why is your face fallen? (Gen. 4:6)

Without awaiting an answer, which I presume God knew and did not want to respond to, He offered His own version of poetic justice likely to befall recalcitrants like Cain:

Surely, if you do right,
There is uplift.
But if you do not do right
Sin couches at the door;
Its urge is toward you,
Yet you can be its master. (Gen. 4:7)

Having issued this warning, God immediately observed the divine consequences of His provocation of Cain:

Cain said to his brother Abel [*Ancient versions*: "Come, let us go into the field"] ... and when they were in the field, Cain set up his brother Abel and killed him. (Gen. 4.8)

Next comes another incredible question from God, "Where is your brother Abel?" (Gen. 4:9), when He seems to know perfectly well that Cain has murdered Abel.[2] Cain's memorable response is less than forthcoming: "I do not know. Am I my brother's keeper?" (Gen. 4.9)

This acerbic answer in the form of a question, I submit, gives us as much insight into Cain's strategic calculations as it does into his shaky morality. First, there seems little doubt that his murder of Abel was premeditated, for he set upon Abel "in the field" (Gen. 4.8), to which, it seems, the brothers journeyed together.[3] Second, warned by God of the presence of sin at his door, Cain cannot be excused for being

---

2. God's question is reminiscent of the rhetorical question He asked Adam in the garden of Eden, "Where are you?" (Gen. 3:9), after Adam had eaten the forbidden fruit and tried to hide from God.

3. A contrary view that Abel's murder was unpremeditated is offered in Sarna (1970, 31).

unaware that his crime might have adverse consequences, even if their exact nature could not be foreseen.

Seething with anger and jealousy over the favoritism shown Abel, and unable to strike out against God directly (even if he had wanted to), Cain did the next best thing—he murdered God's apparent favorite. Under the circumstances, this response to God's taunting from a terribly aggrieved man seems not at all irrational.

Abstracting the choices of God and Cain in such a situation to SB and P, assume that SB can either incite (I) or not incite (Ī) P to become angry or jealous. P, in turn, can rise to the bait and take some untoward action, or try to remain calm and unperturbed, though she may be fuming underneath. I will call P's strategies react (R)—by taking the untoward action—or not react (R̄). I assume the primary and secondary goals of the players are as follows:

SB: (i) wants to incite P; (ii) prefers P to react if incited, not react if not incited.

P: (i) wants SB not to incite her; (ii) prefers to react if incited, not react if not incited.

The resulting game, which is game 19 in the appendix and which I call the Incitement Game, is shown in figure 8.1. Based on SB's primary goal, SB's two best outcomes (4 and 3) are associated with I, and his two worst outcomes (2 and 1) with Ī. Secondarily, SB prefers that P react if she is incited, otherwise not, so 4 and 2 are associated with the main-diagonal entries, and 3 and 1 with the off-diagonal entries. P's two best outcomes (4 and 3) are associated with Ī—just the opposite of SB's—and her two worst outcomes (2 and 1) with I, again the opposite of SB's. But P's secondary goal coincides with SB's: She prefers to react if incited and otherwise not.

The Incitement Game has a unique Nash equilibrium that is also, wherever play starts, its only NME (thus, all states in the anticipation game lead to RI, giving (2,4), whereby P reacts to SB's incitement). This has unfortunate consequences for P (2) compared to SB (4), who obtains his best outcome by successfully provoking P. This is exactly what happened in the Bible when Cain, humiliated by God's treatment of him, struck out against Abel.

SB's choice is reinforced by SB's compellent threat: By choosing I and sticking with it, SB compels P to choose between (1,3) and (2,4). Although the threat outcome, (2,4), is better for both players than

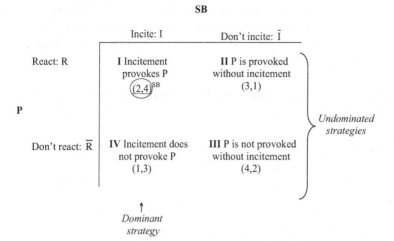

Key: (x,y) = (payoff to SB, payoff to P)
    4 = best; 3 = next-best; 2 = next-worst; 1 = worst
    Nash equilibrium underscored
    Nonmyopic equilibrium (NME) circled
    Superscript SB indicates outcome SB can induce with a compellent threat

Figure 8.1
Outcome and payoff matrix of the Incitement Game (game 19)

the breakdown outcome of (1,3), (2,4) is still an unappetizing outcome
for P.[4]

But at least in the Bible, Cain as P is not without recourse. His reply
to God's questioning him about the murder—that he is not his brother's keeper—can be read as a cleverly constructed challenge to God's
own morality for meddling in the affairs of the brothers.[5] Not that Cain

---

4. I mentioned the use of compellent threats in Chicken in section 5.5 (see note 11 in
chapter 5) and discussed it in the Temptation Game in section 6.3. In Incitement Game
1, there is no evidence that God threatened Cain, beforehand, with harsh punishment if
he murdered Abel. Rather, He simply ignored Cain's offering, which set off a string of
events, following Abel's murder, that I describe in section 8.3.
5. It might also be read as a challenge to God's omniscience—and His complicity—as
suggested by the natural follow-up question posed by Graves and Patai (1963, 58): "Why
should One who watches over all creatures ask one, unless He planned the murder
Himself?" God's role as an accomplice in the murder is also considered, though rejected,
by Wiesel (1977, 58): "Cain could not help but kill: he did not choose the crime; instead
the crime chose him." I find this line of argument, which says that Cain effectively did
not make a choice but was preconditioned to respond, unpersuasive. Cain's response, as
I argued earlier, was not an emotional outburst but instead seems to have been planned.
If this is so, it follows that Cain could anticipate being discovered by God and plan for
his defense.

necessarily knew that God had fomented trouble to test Cain's suscep-
tibility to sin—or simply chose to roil the waters. Whoever was to
blame for Abel's murder, Cain felt deeply wronged and was driven to
take revenge.

8.3   The Blame Game: Cain's Shifting of Blame for Abel's Murder to God

But how does one justify fratricide, and what can one do after the act
to mitigate one's punishment for the crime? I suggest that Cain had
two courses of action open to him once the murder was discovered and
he, irrefutably, was the culprit:

1. Admit to the murder and defend his (D) morality.
2. Admit to the murder and not defend ($\overline{D}$) his morality (e.g., by
asking for forgiveness instead).

True, the first course of action would seem hard to execute shamelessly,
except when it is recalled that the conditions that led to the crime do
not leave God with virtuous intent intact.

Whether one perfidy justifies another, the salient fact is that Cain
did not think his act was unjustified. Even if he had no suspicion of
God's perhaps less-than-virtuous intent in not even acknowledging his
offering, Cain could still defend himself by pleading no responsibility
for his brother's well-being.

Cain's defense is actually more subtle than simply a plea of incul-
pability. He first says that he does not know where his brother is. Could
not this imply that God does, or should, know, and that He bears some
responsibility for Abel, too? The notion that Abel is not Cain's sole
responsibility is then reinforced by Cain's famous question: "Am I my
brother's keeper?" (Gen. 4.9).

This, in my opinion, is a brilliant defense, because it eloquently
contrasts God's responsibility and his own, implicitly suggesting that
there may be questionable morality on both sides. In God's response
to Cain's crime and rhetorical defense, God begins with His own rhetorical question, which He quickly follows with a stiff sentence for a
tiller of the soil:

What have you done? Hark, your brother's blood cries out to me from the
ground! Therefore, you shall be more cursed than the ground, which opened
its mouth to receive your brother's blood from your hand. If you till the soil,
it shall no longer yield its strength to you. You shall become a ceaseless wanderer on earth. (Gen. 4:10–12)

Acting as his own defense attorney, Cain responded to God's sentence with a plea for mercy:

My punishment is too great to bear! Since You have banished me this day from the soil, and I must avoid Your presence and become a restless wanderer on earth—anyone who meets me may kill me! (Gen. 4:13–14)

Note that the crux of Cain's plaintive remonstration is that he might be killed, not that the sentence itself is unjust or inappropriate: It is only "too great to bear." Reminded of this consequence of His sentence, God finds it unpalatable and answers Cain by saying:

"I promise, if anyone kills Cain, sevenfold vengeance shall be taken on him." And the LORD put a mark on Cain, lest anyone who met him should kill him. (Gen. 4:15)

The reason, I believe, that God finds Cain's death unpalatable is because only "a ceaseless wanderer on earth" (Gen. 4:12) can spread far and wide the message of God's retribution for fratricide. If Cain were quickly dispatched, God's great power—and even greater mercy in sparing the murderer's life—would of course not get communicated to the world.

I postulate that God considered two strategies in response to Cain's murder of Abel:

1. Kill (K) Cain.
2. Punish, but not kill ($\overline{K}$), Cain.

If Cain had either admitted his crime or denied it, it is likely that God would probably have chosen to kill Cain. Simply asking for forgiveness would not hold much weight either, because murder, especially of one's brother, is too serious a crime to ignore or sweep under the rug. Moreover, the execution of the murderer would set an impressive precedent.

But there were extenuating circumstances, so punishment short of death must be considered. However, there is no wily serpent, as in the Adam and Eve story (see sections 6.2 and 6.3), that could be implicated and used as exculpation for Cain's sin. The only possible extenuating circumstance was God's complicity—or at least His failure to accept any responsibility for bestirring trouble in the first place, or for not coming to Abel's aid just before his murder (as God had come to Isaac's aid before his sacrifice was consummated).

This failure, and perhaps the resulting guilt God felt, is exactly what Cain's reply to God plays upon. It is as if Cain had said, "He's your responsibility, too; why did you not protect him from my rage, which after all you incited?" If God is not disturbed by this implied question, why would He say that Abel's blood "cries out to *Me* from the ground" (Gen. 4:10; my italics)—not to Cain, not to the world, but to God Himself. God is responsible, too.[6]

God can hardly condemn a man to death when He is also culpable. Consequently, He only banishes Cain from the ground where his brother's blood was shed and spares his life.

In fact, Cain, as I have already indicated, is able to extract still more from a now troubled God: a mark that signals to anyone meeting him that he should not be killed. Would an untroubled God be so attentive to the pleas of a murderer whom He had previously castigated for his crime? Coupled with God's desire to promulgate to the world both His power and mercy, a commutation of Cain's sentence, and making him a marked man, would seem an appropriate punishment.

I model the conflict between Cain and God (Abel is not really a player) as the Blame Game, wherein P and SB must choose between their aforementioned strategies. I postulate their primary and secondary goals to be the following:

SB: (i) wants not to kill P if she defends herself, but to kill P if she does not defend herself; (ii) prefers that P defend herself.

P: (i) wants not to be killed; (ii) prefers not to defend herself if she won't be killed, but to defend herself if she will be killed.

The resulting game, which is game 46 in the appendix, is shown in figure 8.2. Based on P's primary goal, her two best outcomes (4 and 3) are associated with $\overline{K}$, and her two worst outcomes (2 and 1) with K. Secondarily, she does not want to defend herself if she won't be killed (it's a useless expenditure of energy), but this expenditure is worthwhile if she expects to be killed.

SB does not want to kill P if she mounts a credible defense, but otherwise he wants to kill her. Between defending herself and not being killed and not defending herself and being killed, P would prefer the former—saving herself is paramount.

This game has no Nash equilibrium in pure strategies; it cycles without impediments in a clockwise direction. It has two NMEs, with

---

6. Alternatively, one might argue that Abel's blood did not so much elicit shame and self-blame in God as impel Him to avenge Cain's crime. But this interpretation would not explain why God did not respond in kind and simply kill Cain (see subsequent discussion in text).

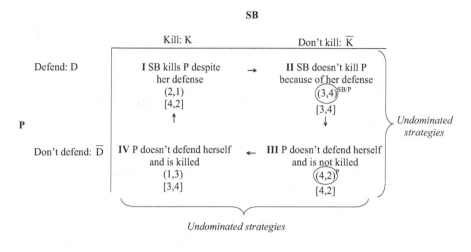

Figure 8.2
Outcome and payoff matrix of the Blame Game (game 46)

(3,4) giving SB her best outcome and (4,2) giving P her best outcome. On the one hand, it is not difficult to show that each player, with moving power, can induce his or her best outcome, making moving power effective. On the other hand, the exercise of threat power by each player leads to (3,4): SB has a deterrent threat of choosing K if P does not agree to (3,4); and P has a compellent threat of choosing D and not budging from it. In each case, the breakdown outcome that the players threaten is (2,1).

Cain, I believe, played the Blame Game masterfully, vigorously defending his morality by shifting responsibility for Abel's murder to God, especially because God admits that "Abel's blood cries out to Me from the ground" (Gen. 4:10). But more than being a guilty party, Cain suggests that God could have interceded to prevent the murder. Furthermore, when Cain questions whether he is his brother's keeper, he implies that his rage—after God paid no heed to his offering—was uncontrollable, diminishing his own responsibility for the murder (it would not be premeditated).

But why did God foment trouble in the first place by not paying heed to Cain's offering? It may have been boredom, as I indicated in section 8.2 and mentioned earlier as a motive for creating a world in which people have free will (section 1.3). But just as important to God might have been His need to signal what kinds of sacrifices He expected from His subjects, and the reverence He believed was due Him. This is a running theme throughout the Hebrew Bible, which we will see more of in the next game.

8.4   The Deception Game: Inducing Saul to Be King, and Then Destroying His Kingship

After the conquest and settlement of Canaan by the Israelites was completed, Samuel became the first prophet and judge of Israel. When the elders of Israel asked him to "appoint a king for us, to govern us like all other nations" (1 Sam. 8:5), Samuel was indignant. So was God, who told Samuel:

Heed the demand of the people in everything they say to you. For it is not you that they have rejected; it is Me they have rejected as their king. Like everything else they have done ever since I brought them out of Egypt to this day—forsaking Me and worshiping other gods—so they are doing to you. Heed their demand; but warn them solemnly, and tell them about the practices of any king who will rule over them. (1 Sam. 8:7–9)

Samuel duly warned the people that the king for whom they importuned him would conscript their sons, take their daughters as "perfumers, cooks, and bakers" (1 Sam. 8:13), seize their choice fields, vineyards, and olive groves and give them to his courtiers, and expropriate a tenth of their grain and vintage. Worst, he said:

you shall become his slaves. The day will come when you cry out because of the king whom you yourselves have chosen; and the LORD will not answer you on that day. (1 Sam. 8:17–18)

But the people refused to listen:

"No," they said. "We must have a king over us, that we may be like all the other nations: Let our king rule over us and go out at our head and fight our battles." (1 Sam. 8:19–20)

So God reluctantly told Samuel, "Heed their demands and appoint a king for them" (1 Sam. 8:22).

At God's behest, Samuel appointed Saul the ruler over the people of Israel, instructing him to

Go down to Gilgal ahead of me, and I will come down to you to present burnt offerings and offer sacrifices of well-being. Wait seven days until I come to you and instruct you what you are to do next. (1 Sam. 10:8)

But Samuel did not arrive after seven days, and "the people began to scatter" (1 Sam. 13:8). To stem their sedition, Saul offered a sacrifice and, just as he had finished, Samuel arrived and rebuked him severely:

You acted foolishly in not keeping the commandments that the LORD your God laid upon you! Otherwise the LORD would have established your dynasty over Israel forever. But now your dynasty will not endure. The LORD will seek out a man after His own heart, and the LORD will appoint him ruler over His people, because you did not abide by what the LORD had commanded you. (1 Sam. 13:13–14)

With his end thus foretold, it becomes clear that Saul has been set up as a fall guy, though previously he had been judged to be the best man for the kingship:

An excellent young man; no one among the Israelites was handsomer than he; he was a head taller than any of the people. (1 Sam. 9:2)

But despite Saul's impressive physical attributes, God could simply not get over the ingratitude of the people. After He had anointed Saul as king, Samuel related God's word and then added his own reprise:

"I brought Israel out of Egypt, and I delivered you from the hands of the Egyptians and all the kingdoms that oppressed you." But today you have rejected your God who delivered you from all your troubles and calamites. For you said, "No, set up a king over us!" (1 Sam. 10:18–19)

Chastened, finally, by God's enmity, the people beseeched Samuel:

Intercede for your servants with the LORD your God that we may not die, for we have added to all our sins wickedness of asking for a king. (1 Sam. 12:19)

Samuel offered the people a reassuring response but concluded with a warning that "if you persist in your wrongdoing, both you and your king shall be swept away" (1 Sam. 12:25).

What makes Saul's case even more pathetic is that he never sought the kingship. He was, by his own self-deprecating testimony, "only a Benjamite, from the smallest of the tribes of Israel, and my clan is the least of all the clans of the tribe of Benjamin!" (1 Sam. 9:21). Previously, Samuel had told Saul that all Israel wanted him. After Saul's disclaimer about his qualifications, he asked, incredulously, "Why do you [Samuel] say such things to me?" (1 Sam. 9:21), especially after Samuel had previously praised Saul's qualifications to the skies.

God did help Saul and the Israelites win victories over the Ammonites and the Philistines. Nonetheless, in addition to the disturbing signs of Saul's eventual ruination alluded to earlier, God held certain things back. For example, when Saul inquired of God whether to pursue the fleeing Philistines after defeating them in battle, he received no answer.

Saul, on the other hand, held nothing back. He made Herculean efforts to pay homage to God, even offering to sacrifice his own son, Jonathan, for breaking a fast he had ordered. But Jonathan, who admitted to the slight impropriety of having "tasted a bit of honey with the tip of the stick in my hand" (1 Sam. 14:43), was saved by the remonstrations of the people.

God, ambivalent up to now about the man whom He had reluctantly invested as king, eventually paved the way for his undoing and final degradation. Samuel, a willing accomplice, related God's word to Saul:

I am exacting the penalty for what Amalek did to Israel, for the assault he made upon them on the road, on their way up from Egypt. Now go, attack Amalek, and proscribe all that belongs to him. Spare no one, but kill alike men and women, infants and sucklings, oxen and sheep, camels and asses. (1 Sam. 15:2–3)

This was a rather harsh edict, especially as several generations had passed since the Israelites had fled Egypt. But God was not known for having a short memory, especially on matters that He took to be personal affronts.

Saul dutifully carried out this savage edict, except to take alive Agag, king of the Amalekites, and spare

the best of the sheep, the oxen, the second born, the lambs, and all else that was of value. They would not proscribe them; they proscribed only what was cheap and worthless. (1 Sam. 15:9)

This sounds like a sensible calculation, except that it contravened God's unsparing edict, causing God to "regret that I made Saul king, for he turned away from Me and has not carried out My commands" (1 Sam. 15:11).

Samuel, whose relations with Saul had never been smooth, was angry, too. His anger was exacerbated when he was told that Saul had set a monument to himself. To make matters even worse, Saul showed no remorse when he greeted Samuel and said he had obeyed God's command. Samuel vehemently disputed this claim, pointing to the sheep and oxen Saul had taken, but Saul retorted that these animals were saved to be given as a sacrifice to God.

Tenaciously, the two men fought back and forth on this issue until Samuel was driven to make God's desires unmistakably clear:

Does the LORD delight in burnt offerings and sacrifices
As much as in obedience to the LORD's command?
Surely, obedience is better than sacrifice,
Compliance than the fat of rams. (1 Sam. 15:22)

Saul, suddenly penitent, then admitted to having sinned because "I was afraid of the troops and I yielded to them" (1 Sam. 15:24).

As I see it, the game God/Samuel played against Saul was an unfair one from the start. Saul, as it were, was plucked from nowhere and set up as king, much as Moses earlier was catapulted into the position of leader of the Israelites. But there is a crucial difference between these figures: On the one hand, God was willing to give Moses practically all the help he wanted, including a brother who stood at his side and spoke his lines for him (see section 9.2); on the other hand, Saul received no help from God, who apparently was very upset that His chosen people would so much as wish for a king, much less demand one. Feeling utterly dejected by such inconsiderateness, God found Saul an obvious target on whom to vent His anger.

Because of their strained relationship, Samuel was not much help to Saul, either, though Samuel's role as judge should have made him an ideal coach to a neophyte king. However, Samuel had his own grudge to bear: His sons had been rejected by the elders of Israel even though, in his old age, Samuel had opportunistically tried to appoint them as judges to replace him. The elders had told Samuel simply that his sons "have not followed your [Samuel's] ways" (1 Sam. 8:5), but the Bible also reports that the sons "were bent on gain, they accepted bribes, and subverted justice" (1 Sam. 8:3).

It is certainly not surprising that a dejected God and His embittered sycophant, Samuel, would lash out at a hapless Saul. What is surprising, perhaps, is that they kept their vendetta muted at first, and mostly used innuendos. They even went through the motions of supporting the king they could not stand, lest they be accused of reneging on their commitment to respect the people's desire.

In fact, it is "the people," unwittingly, who probably play the most significant role in Saul's rise and fall. When they impertinently cry out for a king, God reluctantly accedes to their wish; when they recant—

partially to appease God—He seizes upon their reversion. Whether God in these instances is a democrat or a fawning autocrat I hesitate to say.

For all the vaunted respect the people receive, they do not give very substantial reasons for their initial wish—just a desire to be like everybody else, with a king to lead them in battles. No wonder God felt resentment that His chosen people were not being very discriminating about their leadership preferences!

But when the people admit to having gone too far in their request for a king and evince a willingness to retract it, God and Samuel are ready to step in and engineer Saul's fall from favor. Saul's failure to carry out to the letter God's command to eviscerate the Amalekites in a horrific slaughter is just the excuse Samuel needs to justify an end to Saul's reign:

Because you rejected the LORD's command,
He has rejected you as king. (1 Sam. 15:23)

Why did not Saul strictly obey God's command? Once Saul had learned from Samuel that God was no longer on his side (1 Sam. 13:13–14), he may have decided he had no alternative but to fend for himself—and to listen to the people (1 Sam. 15:24) at least as much as to God. In this manner, God's abandonment of Saul may have diminished Saul's faith, which in turn triggered his disobedience and eventual downfall, thereby fulfilling Samuel's earlier prophecy.

As a model of the conflict between God/Samuel and Saul, I assume in the Deception Game that SB can either support (S) or not support (S̄) P. P, in turn, can obey (O) or not obey (Ō) SB's command to offer a sacrifice. The twist is that, in the biblical game, the support that God/Samuel give Saul is meant to appease the people's desire for a king and lead Saul to believe that they do so. In the Deception Game, P, believing that SB's support is genuine and lasting, is less than meticulous about preparing the sacrifice, withholding certain items for her subjects.

I assume that the primary and secondary goals of the players are as follows:

SB: (i) wants not to support—even to undermine—P; (ii) prefers to support P when she is obedient, but not to support her when she is disobedient.

P: (i) wants to obey SB if he supports her, but disobey him if he does not; (ii) prefers obedience, whether SB supports her or not.

P's secondary goal may sound odd: Why would P prefer obedience if SB does not support her? If she is obedient, she can at least claim that she made an effort to gain SB's support, whereas without trying, she would never know whether she might have succeeded.

The game that follows from these goals is game 17 in the appendix, which is shown in figure 8.3. SB has a dominant strategy of not supporting P, which leads to the unique Nash equilibrium, (3,4), that is the only NME in the Deception Game, wherever play starts.

SB's choice is reinforced by his compellent threat: By choosing S̄ and sticking with it, SB compels P to choose between (2,3) and (3,4). (P also has a compellent threat of choosing Ō that reinforces (3,4), but this is not P's best outcome.) Although the threat outcome, (3,4), is better for both players than the breakdown outcome, (2,3), P would still prefer (4,2), when she is obedient and SB supports her. Unfortunately for P, this outcome is not a rational choice for the players, because SB has no reason to support her and, in turn, she has no reason to offer a bountiful sacrifice and ignore the needs of her subjects, who were her original supporters.

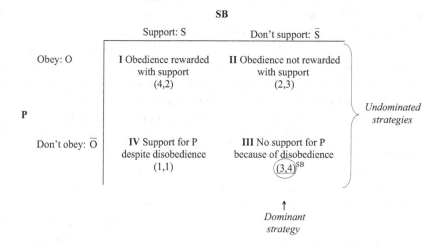

Key: (x,y) = (payoff to SB, payoff to P)
    4 = best; 3 = next-best; 2 = next-worst; 1 = worst
    Nash equilibrium underscored
    Nonmyopic equilibrium (NME) circled
    Superscript SB indicates outcome SB can induce with a compellent threat

Figure 8.3
Outcome and payoff matrix of the Deception Game (game 17)

To conclude Saul's sad story, the final rupture between him and God is echoed, symbolically, in the parting scene between Samuel and Saul:

As Samuel turned to leave, Saul seized the corner of his robe, and it tore. And Samuel said to him, "The LORD has this day torn the kingship over Israel away from you and has given it to another who is worthier than you. Moreover, the Glory of Israel does not deceive or change His mind, for He is not human that He should change His mind." (1 Sam. 15:27–29)

There are several ironies in this passage, beginning with the fact that Samuel's statement does not square with the facts of Saul's reign and the events that preceded it. God *did* change His mind about a king when the people clamored for one. Also, once Saul was installed, the divine support he had received quickly evaporated, which casts doubt on the statement that God "does not deceive." Moreover, when the people themselves had second thoughts about having a king after the disadvantages were pointed out to them, God was more than willing to escalate innuendos into a complete reversal of His original position. Although Saul certainly showed signs of weakness in deferring to the people and not strictly obeying God's commands, his degradation by God/Samuel has devious undertones that smack of a frame-up.

God and Samuel's actions in this sordid affair are, I think, entirely consistent with their motives, which essentially involve avenging the insult that the people handed to them by demanding a king. (Why the people are not punished directly for their apostasy is not evident, but presumably God/Samuel find it easier to vent their frustration and anger on the living embodiment of their displeasure.) Buffeted by their increasingly aggressive and duplicitous tactics, the luckless Saul becomes a helpless pawn, and his kingship turns sour.

I find it hard not to feel sympathy for Saul as he tries to stave off defeat in an unfair game against inimical forces beyond his control. One should not overlook the fact, however, that if God and Samuel behaved ignominiously in this affair, the fickle people collectively stand behind them, influencing their choices as well as Saul's: They are the significant behind-the-scenes player in this tragic story.

8.5 Conclusions

Both game theory and TOM explain well the outcomes in two of the three games described in this chapter (Incitement and Deception). The unique Nash equilibria and NMEs in each of these games coincide, which can be reinforced by a compellent threat by SB in the Incitement Game, and by both players in the Deception Game. In the biblical stories that inspired these games, the players made the choices predicted by game theory and TOM.

There is no Nash equilibrium in pure strategies in the Blame Game, but there is a unique NME that both moving and threat power by either player can induce. This NME, too, was the choice of the players in the biblical game, though there is a second NME that Saul's moving power (if he had it) induces.

While moving or threat power may assist a player in the implementation of these outcomes, the fact that NMEs were chosen in all three games suggests that SB's superiority did not have a decisive influence. If SB had been an ordinary player, the same outcomes would probably have occurred, so the choice of these outcomes does not single out SB as necessarily superior.

But the undecidability of SB in these games is not the factor that most distinguishes them. It is the apparent immorality of SB's actions in each game that is distinctive and also troubling:

- In the Incitement Game, God lights the fuse that sets off Cain's murder of Abel.
- In the Blame Game, God's seeming admission of partial responsibility, and perhaps feelings of guilt, for Abel's murder leads Him to step back from killing Cain. Marking Cain enables God to signal His extreme displeasure with Abel's murder while allowing Cain to live.
- In the Deception Game, God and Samuel set up Saul to be king—deceiving him into thinking their support is heartfelt—but then they conspire to destroy his kingdom, which they succeed in doing after planting seeds of doubt in his subjects.

In my opinion, God's actions in each game are hard to defend on ethical grounds. Consequently, I believe He must bear some responsibility for creating evil. To be sure, SB, based on his goals, has his reasons for acting as he does, as does P, whose own actions certainly are not always praiseworthy.

True, rational choices need not satisfy an ethical standard, either for SB or P. But this makes SB difficult to distinguish from P when judged not just by the rationality of his choices but also by their morality, which exacerbates the undecidability problem.

9 The Defiance (Manipulated), Pursuit, and Salvation Games

9.1 Introduction

All of the games in this chapter are based on conflicts that Moses and God had with Pharaoh and, later, with the Israelites themselves. Beginning with the birth of Moses and his rescue and adoption by Pharaoh's daughter as an infant, Moses had a conflicted identity. As an adult, Moses takes the side of God in an extended struggle with Pharaoh over the fate of Egypt's Israelite slaves. After the demise of Pharaoh, Moses is preoccupied with keeping his people appeased, and spared from the wrath of God, as they wander in the wilderness for forty years.

Moses's leadership problems are complicated in the beginning by his relationship with God—and his insecurity in their relationship—and then by a growing independence and turning away from God. Although Moses was never outrightly disloyal to God in his later years, he had constantly to suffer God's indignation with His unruly subjects, the Israelites. That Moses was largely successful in mollifying God attests both to his adroitness and God's perspicaciousness in designating him as leader, following God's choice of Abraham to be the founder of a promised great nation (see section 7.2).

As the stories I describe in this chapter attest, God and Moses complement each other beautifully: As a foil for God's rashness, Moses is superb; as a user of Moses's conciliatory talents, God is exceptional. But more than being a good judge of character, God is able to visualize and then implement game scenarios—first with an obstinate Pharaoh (Defiance [Manipulated] Game and Pursuit Game) and then with the rebellious Israelites (Salvation Game)—that display His great powers.

While suffering some setbacks, God seems to have enjoyed rising to the challenges He faced from varied opponents. Indeed, He admits that He wanted to promulgate, even glorify, His might, as did SB against P in the Constraint Game (with Adam and Eve) and in the Incitement Game (with Cain).

I next analyze three games between SB and P, in which Pharaoh is P in two games and Moses is P in the third:

1. In the Defiance (Manipulated) Game, God and Moses repeatedly raise the stakes against a defiant Pharaoh, but unlike previous games I have analyzed, Pharaoh is hobbled by the fact that God "stiffens" or "hardens" his heart. This robs Pharaoh of any decision-making ability, preventing him from being a player who can make independent choices and turning the Defiance (Manipulated) Game into a decision in which God calls the shots (though I later discuss an alternative interpretation).

2. In the Pursuit Game, Pharaoh regains at least some of his ability to make decisions. Realizing that Egypt will be crippled by the loss of its slaves, Pharaoh decides to pursue the fleeing Israelites. When God steps in to stop the Egyptian forces, He once again asserts His power to frustrate Pharaoh.

3. In the Salvation Game, God faces a greater challenge, because it is His chosen people, the Israelites, who defy both Him and Moses by building a molten (golden) calf to worship. But Moses engineers a solution, whereby he intervenes with God in the Salvation Game while forcing many Israelites to pay a high price for their idolatry.

God is invincible in the first two games, but He is not beyond persuasion in the Salvation Game. Because a player without His powers almost surely would not have succeeded against Pharaoh, there is not a decidability problem in the Defiance (Manipulated) Game and the Pursuit Game. As SB, however, God's superiority in the Salvation Game is more ambiguous, because Moses succeeds in persuading God not to wipe out the Israelites entirely and start all over again.

As in earlier stories, however, God does not heap distinction upon Himself by the methods He uses to demolish His opponents. This renders His moral superiority questionable, especially in the Defiance (Manipulated) Game, whereby Pharaoh becomes a helpless pawn who, at God's behest, continually raises the stakes.

Although Moses starts out as insecure and wimpish, he evolves into a forceful advocate of his people. To understand Moses's development in his formative years—and God's role in setting up Moses to play a leadership role—it is instructive to look at Moses's origins and early life in Egypt.

9.2 Enter Moses

Jacob, the son of Isaac and the grandson of Abraham, triumphed over his older brother Esau in obtaining the birthright of his father when his mother, Rebekah, duped Isaac into believing Jacob was Esau. Following a conflict between Jacob's son, Joseph, and Joseph's brothers (Brams [1980] 2007, 66–75), the next generation of Israelites "were fertile and prolific; they multiplied and increased very greatly so that the land was filled with them" (Exod. 1:7). However, at this time there was an unfortunate development:

a new king arose over Egypt, who did not know Joseph. And he said to his people, "Look, the Israelite people are much too numerous for us. Let us deal shrewdly with them, so that they may not increase; otherwise in the event of war they may join our enemies in fighting against us and rise from the ground." (Exod. 1:8–10)

To deal with a possible insurrection by the Israelites, the Egyptians

set taskmasters over them to oppress them with forced labor; and they built garrison cites for Pharaoh: Pithom and Ramses. But the more they were oppressed, the more they increased and spread out, so that the [Egyptians] came to dread the Israelites. (Exod. 1:11–12)

The Egyptians became even more cruel and ruthless, but to no avail. In desperation, Pharaoh decided that truly draconian measures were needed, so he ordered midwives to kill newborn boys, presumably because females could better be assimilated into the Egyptian population, while the males posed the threat of overthrow (Ackerman 1974, 85). Being God-fearing women, however, the midwives disobeyed Pharaoh's order, under the pretext that Hebrew women were more vigorous than Egyptian women and, while in labor, gave birth before the midwives could get to them.

Frustrated, Pharaoh then ordered that every newborn Hebrew boy be thrown into the Nile. But this sacrifice was unbearable for a certain Levite woman who bore a son:

When she saw how beautiful he was, she hid him for three months. When she could hide him no longer, she got a wicker basket for him and calked it with bitumen and pitch. She put the child into it and placed it among the reeds by the bank of the Nile. And his [the child's] sister stationed herself at a distance, to learn what would befall him.

The daughter of Pharaoh came down to bathe in the Nile, while her maidens walked along the Nile. She spied the basket among the reeds and sent her slave girl to fetch it. When she opened it, she saw that it was child, a boy crying. She took pity on it and said, "This must be a Hebrew child." (Exod. 2:2–6)

Happily, when the boy's sister asked Pharaoh's daughter if she would like assistance, she said yes; the sister then sought out her mother, standing nearby, to care for her son. Pharaoh's daughter adopted the boy and called him Moses.

This touching story not only illustrates the ambivalence within Pharaoh's household but, as a literary device, works well to compound the irony in the next episode in Moses's life. Now fully grown, Moses observes an Egyptian strike a Hebrew:

He [Moses] turned this way and that and, seeing no one about, he struck down the Egyptian and hid him in the sand. When he went out the next day, he found two Hebrews fighting; so he said to the offender, "Why do you strike your fellow?" He retorted, "Who made you chief and ruler over us! Do you mean to kill me as you killed the Egyptian?" (Exod. 2:12–14)

"Frightened" (Exod. 2:14) by this response, Moses, who had thought that his murder of the Egyptian had gone undetected, quickly made a sagacious calculation—"Then the matter is known!" (Exod. 2:14)—and took precautions.

Years passed, the old Pharaoh died, but

the Israelites were groaning under the bondage and cried out; and their cry for help from the bondage rose up to God. God heard their moaning, and God remembered His covenant with Abraham and Isaac and Jacob. God looked upon the Israelites, and God took notice of them. (Exod. 2:23–25)

God then singled out Moses for the task of delivering the Israelites from slavery, but Moses needed considerable convincing. Patiently, God offered various reassurances: He recreated Himself in a burning bush not consumed by fire; He told a reluctant Moses, "I will be with you" (Exod. 3:12) and "I have taken note of you and of what is being done to you in Egypt" (Exod. 3:16); and He provided Moses with detailed instructions on what to say to Pharaoh when he saw him.

But Moses still could not believe that he was up to the task. So for good measure God performed a few more miracles that He told Moses he himself could reenact to convince Pharaoh—as well as his own people—that he was indeed God's duly appointed representative. God even assuaged Moses's fear of not being "a man of words" (Exod. 4:10) by saying to him:

Who gives man speech? Who makes him dumb or deaf, seeing or blind? Is it not I, the Lord? Now go, and I will be with you as you speak and will instruct you what to say. (Exod. 4:11–12)

When Moses still protested his rhetorical inadequacies, God lost His patience and grew angry. Finally, He proposed a pragmatic solution:

There is your brother Aaron the Levite. He, I know, speaks readily. Even now he is setting out to meet you, and he will be happy to see you. You shall speak to him and put the words in his mouth—I will be with you and with him as you speak, and tell both of you what to do—and he shall speak for you to the people. Thus he shall serve as your spokesman, with you playing the role of God to him. And take with you this rod, with which you shall perform the signs. (Exod. 4:14–17)

The portrait of Moses that emerges is that of a meek, insecure man suffering almost debilitating doubts about his strength of character and physical capabilities. He is a constant worrier and continually seeks reassurance. I judge Moses at this point in his career to be a vacillating figure with an almost morbid aversion to becoming entrapped in a situation that entails substantial risk. Basically, he seems inept.[1]

9.3   The Defiance (Manipulated) Game: Pharaoh and the Ten Plagues

In the end, God's entreaties to Moses, and His resounding demonstrations of miraculous powers, succeed in allaying Moses's doubt and fears. But before sending Moses on his way, God still finds it necessary to give a final reminder to His anxious servant:

When you return to Egypt, see that you perform before Pharaoh all the miracles that I have put within your power. (Exod. 4:21)

This includes taking a rod that God had given Moses and turning it into a snake and then back again.

If Moses is a pusillanimous figure who constantly needs reassurances and reminders, the same cannot be said for God. He seems to crave confrontation. Moreover, He is not above manipulating a situation so as to heighten the dramatic tension. After telling Moses to display all his marvels before Pharaoh, for example, He reveals a

---

1. On the other hand, Moses was anything but inept when he killed the Egyptian and then calculated how to escape from being caught and facing a murder charge. Moses's true character is hard to decipher at this point, but the picture becomes clearer later.

devious strategy for escalating the conflict with Pharaoh: "I, however, will stiffen his heart so that he will not let the people go" (Exod. 4:21).

This is blatant manipulation. Earlier I argued (see section 8.2) that God had provoked Cain into murdering his brother by accepting Abel's offering while refusing Cain's, thereby implanting irrepressible jealousy in Cain. But that was manipulation of the physical circumstances of the situation. Now God seems ready to practice some mind manipulation as well, contrary to the notion that humans have free will (see section 1.3).

I have no explanation for this new-found ability of God, though His reasons for using it seem rather clear (to be discussed shortly). In fact, the mind manipulation of Pharaoh is one of the few instances in the Hebrew Bible in which a human character is robbed of his or her free will and, as such, the ability to make his or her own choices (for another instance, see Josh. 11:20).

To be sure, Pharaoh in the end does change his mind and seems to act on his own volition, so God's control appears not to be total. Also, Moses, spokesman for God that he is in the beginning, later asserts his independence, so he, too, emerges as more than just a mouthpiece of God. Nonetheless, it must be stressed, as in the Cain and Abel story, that the deck is heavily stacked by God at the outset—and only later do His antagonists emerge as full-fledged game players.

God signals that He is in charge, and beyond the control of any other forces, when He offers the following elliptic response to Moses's question of what to tell the Israelites if they ask God's name (quoted earlier in the preface): "Ehyeh-Asher-Ehyeh [I Am That I Am]" (Exod. 3:14). God amplifies this cryptic statement by saying:

Thus shall you say to the Israelites, "Ehyeh [I Am] sent me to you." And God said further to Moses, "Thus shall you speak to the Israelites: The LORD, the God of your fathers, the God of Abraham, the God of Isaac, and the God of Jacob, has sent Me to you:
This shall be My name forever,
This My appellation for all eternity." (Exod. 3:14–15)

If the legitimate bases of God's authority are not exactly transparent from His "I Am That I Am" statement, then the next phase of the story offers some clues. While God may like to turn aside existential questions like Moses's with enigmatic responses, He is not so successful in covering up His role in human affairs when He is called upon to act.

This point is well illustrated after Moses and Aaron approach Pharaoh, and Moses entreats him with the words of "the LORD, the God of Israel" (Exod. 5:1):

Let My people go so that they may celebrate a festival for Me in the wilderness. (Exod. 5:1)

Pharaoh's mocking reply is:

Who is the LORD that I should heed Him and let Israel go? I do not know the LORD, nor will I let Israel go. (Exod. 5:2)

Pharaoh then follows up his reply by ordering that the Israelites no longer be provided with straw for making bricks but that the daily quota of bricks to be produced remain the same. When the quotas are not met, the Israelite foremen are beaten, and in a rage Pharaoh accuses them of shirking.

Upon leaving Pharaoh, the maligned foremen meet Moses and Aaron, who are waiting for them. They curse their ostensible benefactors and charge them with betrayal:

May the LORD look upon you and punish you for making us loathsome to Pharaoh and his courtiers—putting a sword in their hands to slay us. (Exod. 5:21)

Moses and Aaron indeed would seem to have good reason to question the perverse effects of their actions.

With the situation looking bleak, Moses implores God:

O Lord, why did You bring harm upon this people? Why did You send me? Ever since I came to Pharaoh to speak in Your name, he has dealt worse with this people; and still You have not delivered Your people. (Exod. 5:22–23)

God's answer now reveals more of His grand design:

You shall soon see what I will do to Pharaoh: he shall let them go because of a greater might; indeed, because of a greater might he shall drive them from his land. (Exod. 6:1)

In fact, a harsh Pharaoh and counterproductive appeals to him by Moses and Aaron are but preliminaries to the main event.

Before this event commences, the Bible offers a key insight into God's rationale for staging the preliminaries that I have described. After repeating His earlier pledge to renew His covenant with Abraham, Isaac, and Jacob, and then reaffirming His pledge to rescue the Israelites from slavery, God speaks of His own role in the rescue operation:

I am the LORD. I will free you from the burdens of the Egyptians and deliver
you from their bondage. I will redeem you with an outstretched arm and
through extraordinary chastisements. And I will take you to be My people, and
I will be your God. And you shall know that I, the LORD, am your God who
freed you from the labors of the Egyptians. (Exod. 6:6–7)

Note the connection between the subjects of these two verses—six
first-person references, modified by God's self-identification twice as
"the LORD" and twice as "God"—with the verbs "free," "deliver," and
"redeem." Not only can there be no doubt about who is doing all these
things, but God also does not hide His opinion that His chastisements
will be "extraordinary," that His efforts "you shall know." If His place
in the world were not already apparent, God concludes that next verse
with a final "I the LORD" (Exod. 6:8).

Why does God so relentlessly hammer away at His self-designated
role as savior of the Israelites? I believe there are at least two reasons.
First, He probably anticipated—not without reason—future trouble
with the people He has chosen to adopt; and the previously quoted
statements help to establish His magnanimity when they are in desperate straits.

Second, and more immediate, God, as I indicated earlier, continually
seeks praise and admiration, and what better way is there to achieve
it than to "free … deliver … redeem" a desperate people? Clearly, the
salvation of the Israelites becomes more praiseworthy and admirable
the more trying their condition is. Unwittingly, Moses and Aaron succeeded in aggravating it. It appears all in God's design first to exacer-
bate an already deplorable situation for His chosen people and then to
come gallantly to their rescue.

At this point, Pharaoh does not recognize the consequences of confronting God. He also does not realize that God's objective is not just
the release of the slaves, whom Pharaoh especially feared after brutal-
izing them, but that God also wants to magnify His achievements to
the nth degree.

There is one troublesome question in this interpretation. Except for
reasons of deep insecurity, it is not apparent why God does not let His
great deeds speak for themselves instead of running on ad nauseam
about His accomplishments. The explanation, I believe, is one alluded
to earlier: Anticipating that the Israelites will later deviate, He wants
to get His "extraordinary chastisements" (Exod. 6:6) on the record.

Whatever the reasons for God's unabashed insistence on His power
and glory, it is evident that He sets the stage for a climatic confrontation

that He thinks will enhance His image. I next explore how God orchestrates events to achieve this objective.

First, He makes unmistakably clear that He alone is the LORD, and that He also determines the rules of the game. That He is the only consequential player in this game is made manifest in an alternative account of the instructions God gives Moses before he and Aaron meet Pharaoh:

> See, I place you in the role of God to Pharaoh, with your brother Aaron as your prophet. You shall repeat all that I command you, and your brother Aaron shall speak to Pharaoh to let the Israelites depart from his land. But I will harden Pharaoh's heart, that I may multiply My signs and marvels in the land of Egypt. When Pharaoh does not heed you, I will lay My hand upon Egypt and deliver My ranks, My people the Israelites, from the land of Egypt with extraordinary chastisements." (Exod. 7:1–4)

Note the repetition of the self-laudatory phrase "with extraordinary chastisements."

When Pharaoh proves "stubborn" (Exod. 7:14)—despite a demonstration by Aaron, at Moses's insistence, of his ability to turn his staff into a serpent that then swallows up serpents conjured up by Pharaoh's magicians—God escalates the conflict. Progressively, through Moses and Aaron, He brings harsher and harsher plagues down upon the Egyptians. Although Egyptian magicians, through their spells, succeed in duplicating the first two plagues—turning the Nile into blood and infesting the land with frogs—after the second plague Pharaoh acknowledges God's existence and gains relief from the frogs only by promising to let the Israelites go and give sacrifice to their God.

Pharaoh retracts his pledge, however, and becomes once again obstinate. Successive plagues of lice, swarms of insects, pestilence against animals, dust that produces boils, thunder and hail, locusts, and darkness, which Pharaoh's magicians are no longer able to reproduce, each time inducing Pharaoh to yield, but only temporarily. Then Pharaoh becomes implacable again—as God predicted he would—and thereby brings on himself and the Egyptians a new and more ghastly plague.

The hail of the seventh plague, which levels everything in the open, destroying "man and beast and all the grasses of the field" (Exod. 9:22), strikes the fear of God even in some of Pharaoh's courtiers. This is the first sign that their sufferance is not inexhaustible. The darkness of the ninth plague "that can be touched" (Exod. 10:21)—implying that even breathing becomes difficult in the palpable gloom—so enrages Pharaoh that he threatens Moses with death:

Be gone from me! Take care not to see me again, for the moment you look upon my face you shall die. (Exod. 10:28)

Moses responds by signaling that the climax is near:

You have spoken rightly. I shall not see your face again! (Exod. 10:29)

The coup de grâce comes with the tenth plague, which God tells Moses will be the last:

Toward midnight I will go forth among the Egyptians, and every first-born in the land of Egypt shall die, from the first-born of Pharaoh who sits on his throne to the first-born of the slave girl who is behind the millstones; and all the first-born of the cattle. And there shall be a loud cry in all the land of Egypt, such as has never been or will ever be again; but not a dog shall snarl at any of the Israelites, at man or beast—in order that you may know that the LORD makes a distinction between Egypt and Israel. (Exod. 11:4–7)

True to His word, "there was a loud cry in Egypt, for there was no house where there was not someone dead" (Exod. 12:30). Pharaoh could withstand the devastating effects of the plagues no longer and summoned Moses and Aaron:

Up, depart from among my people, you and the Israelites with you! Go, worship the LORD as you said! Take also your flocks and your herds, as you said, and begone! And may you bring blessing upon me also! (Exod. 12:31–32)

A fitting testimony to the sway God finally holds over even His enemies.

But what, exactly, is the game that God is playing with Pharaoh? Earlier He played a kind of assurance game with Moses to bolster the latter's courage and resolve before the confrontation. Then, after Moses killed an Egyptian and aggravated the situation for the Israelites, God offered reassurances. In the climactic struggle with Pharaoh that ensues, the question is why did God unleash the series of plagues that He did? After all, if He had full control over Pharaoh's mind, would it not have been easier and more expeditious for Him to make Pharaoh yield quickly, rather than endure the horror of ten hideous plagues?

The reason God desired a protracted conflict is hinted at in His several boastful statements about acting as savior of the Israelites in their time of crisis (which God, of course, had helped to deepen and prolong). But God also had another purpose in mind, which He expressed to Moses and Aaron just prior to their audience with Pharaoh (the first part of God's instructions were quoted earlier):

And the Egyptian shall know that I am the Lord, when I stretch out My hand over Egypt and bring out the Israelites from their midst. (Exod. 7:5)

In other words, God did not just want to save His chosen people, which He could have done well enough by making Pharaoh less stubborn and more compassionate. He also wanted to teach Pharaoh and the Egyptians an unforgettable lesson, which He seems to have succeeded in doing after the tenth plague. Indeed, not only did Pharaoh ask for God's blessing, as I indicated earlier, but also the Egyptians themselves hastened the exodus:

The Egyptians urged the people [Israelites] on, to make them leave in haste, for they said, "We shall all be dead." … And the LORD had disposed the Egyptians favorably toward the people and they let them have their request; thus they stripped the Egyptians. (Exod. 12:33–36)

So were the tables turned on the Egyptians, again through a little mind manipulation that "disposed [them] favorably."

This manipulation of people, especially of Pharaoh, renders his confrontation with Moses/God less of a game than a decision (see chapter 2), as shown in figure 9.1 for P and SB (I have called it a game, because it starts off as a two-person game before it becomes a one-person game when God hardens Pharaoh's heart). SB's two strategy choices are to stop or to continue the plagues. P, by contrast, does not have strategy choices since SB controls her behavior; she can better be thought of as a state of nature, in which her two possible states are to be "submissive" or "defiant (up to a point)."

These states do not arise strictly by chance, however, as is assumed in decision theory; they are themselves chosen by SB. Thus, SB in effect

State of nature

		P submissive	P defiant (up to a point)	
SB	Stop plagues	SB's power not revealed (3)	SB powerless (1)	Undominated strategies
	Don't stop plagues	SB harsh (2)	SB's power revealed (4)	

Key: 4 = best; 3 = next-best; 2 = next-worst; 1 = worst

Figure 9.1
The Defiance (Manipulated) Game as a decision

chooses both a strategy—to stop or to continue the plagues—and a state of nature—to make P submissive or defiant (up to a point). This is a violation of the assumption that states arise independently of a player's choice, which I discussed in connection with Pascal's wager and the Search Decision (chapter 2, note 3),

The outcomes of each pair of SB's choices (i.e., of one strategy and one state of nature) are given in figure 9.1. If SB stops the plagues, on the one hand, he obtains his next-best outcome (3) when P cooperates by being submissive, but his worst outcome (1) when P remains defiant, which would make SB appear powerless to stop a determined opponent.

By continuing the plagues, on the other hand, SB appears unnecessarily harsh if P is submissive, which is his next-worst outcome (2). But if P is defiant (up to a point), SB obtains his best outcome (4) by demonstrating his ability to inflict great harm and, in the end, defeat his foes.

In the Bible, God (SB) taught Pharaoh (P) a painful lesson by making the appropriate row and column choices in figure 9.1 (second row, second column) to obtain SB's best outcome (4).[2] From the passages I quoted earlier, it seems to have been taken to heart by both Pharaoh and his Egyptian subjects.

But the game with Pharaoh is not over, because Pharaoh soon has a change of heart and emerges as a new player, seemingly now out of God's mental control. This is a genuine two-player game, which I set up by describing what happens after the Israelites flee Egypt.

9.4   The Pursuit Game: Pharaoh and the Israelites in the Wilderness

In the biblical version of the Defiance (Manipulated) Game, God held all the cards; there was no question that He was the demonstrably superior player, at least in forcing Pharaoh to let the Israelites flee from Egypt. But the story can also be read as one in which Pharaoh eventually reached the point where dogged defiance appeared futile, or at least no better than suicide in installments. Under this interpretation, Pharaoh, though rendered obstinate by God, in the end asserted his

---

2. Notice that God does not have a dominant strategy of stopping or continuing the plagues—it depends on what state of nature obtains. But because God also chooses the state of nature, this lack of dominance does not prevent SB from ensuring his best outcome when Pharaoh capitulates.

release from God's control and made the rational calculation that it was in his own best interest to let the Israelites go.

This is why in figure 9.1 I have attributed to SB (God) the power to make P (Pharaoh) "defiant," but only "up to a point." SB's mastery of the situation loses its force when P caves in, which is an interpretation more consistent with P's possessing free will and making her own prudent choice.

But does Pharaoh act prudently, or is it part of God's larger design to have Pharaoh suffer a temporary setback only to reemerge as an even more defiant antagonist whom God can then stamp out once and for all, without fear of being condemned as merciless? The Bible seems contradictory on this point. On the one hand, as the Israelites flee Egypt, God manipulates the situation so that

Pharaoh will say of the Israelites, "They are astray in the land; the wilderness has closed in on them. Then I will stiffen Pharaoh's heart and he will pursue them, that I may assert My authority against Pharaoh and all his host; and the Egyptians shall know that I am the LORD. (Exod. 14:3–4)

On the other hand, in the next verse, the Bible says that Pharaoh had good reason, apart from observing the Israelites in difficult country, to pursue them:

When the king of Egypt was told that the people had fled, Pharaoh and his courtiers had a change of heart about the people and said, "What is this we have done, releasing Israel from our service?" (Exod. 14:5)

Not knowing whether or not Pharaoh is a tool of God, I propose to look at the Pursuit Game in two ways. First, consider the game as if P were an independent player, who must decide whether to pursue or not pursue her adversary, with whom SB sides. SB must decide whether to help or not help P's adversary. I assume the primary and secondary goals of the players are as follows:

SB: (i) wants to help P's adversary if pursued, but not to help if not pursued; (ii) prefers to help P's adversary, whether pursued or not.

P: (i) wants SB not to help if P's adversary is pursued, but to help if not pursued; (ii) prefers to pursue P's adversary, whether helped or not.

The payoffs implied by these goals are given below the brief verbal descriptions of each outcome in figure 9.2. I justify the players' rankings next.

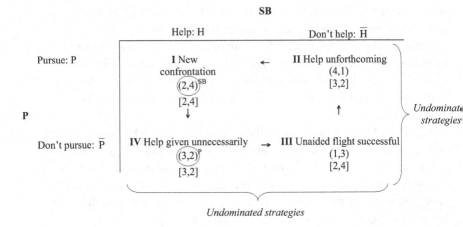

Key: (x,y) = (payoff to SB, payoff to P)
      4 = best; 3 = next-best; 2 = next-worst; 1 = worst
      Anticipation game given in brackets
      Nonmyopic equilibrium (NME) circled
      Arrows indicate progression of states in strongly cyclic game
      Superscripts P and SB indicate outcomes each player can induce with moving
          power

Figure 9.2
Outcome and payoff matrix of the Pursuit Game (game 42)

A new confrontation is next worst (2) for Pharaoh, for even though the loss of the Israelite slaves would be catastrophic for Egypt, Pharaoh risks a great deal in a new confrontation with God/Moses. This is the best outcome (4) for God/Moses, because God, especially, relishes the opportunity to display His power. By contrast, if help is unforthcoming, Pharaoh would have an easy victory over the unarmed Israelites, leading to his best outcome and the worst for God/Moses at (4,1).

If Pharaoh does not pursue the Israelites when they are protected in their escape, he does not suffer defeat, which is good (3), but there is hardly any need for God/Moses to protect them, which is not good (2). Finally, it would be unthinkable for Pharaoh not to pursue the Israelites if they are unprotected (1), while for God/Moses it would be a satisfactory outcome (3) since the Israelites do not need help. However, there would not be the confrontation that God seeks not only to vanquish Pharaoh but also to further enhance His reputation for protecting His chosen people.

Although Pharaoh makes his choice without knowing what course of action God/Moses will choose, God/Moses learn quickly that Pharaoh is in hot pursuit. This suggests that a 2 × 4 matrix, in which God/Moses make their choice contingent on Pharaoh's prior choice, is a better representation of the Pursuit Game (see section 6.2 for an example). In fact, (2,4) is the unique Nash equilibrium in the 2 × 4 game, whereas the 2 × 2 game cycles in a clockwise direction and has no Nash equilibrium in pure strategies (see figure 9.2).

This is game 42 in the appendix, and it has two NMEs, (2,4) and (3,2); which one will be chosen depends on where play starts, as shown in the anticipation game. If there is a new confrontation, as the Bible reports occurred, then the players will have no reason to depart from this state, yielding (2,4) if they think nonmyopically. Furthermore, this outcome can be induced if SB has moving power, which is plausible, whereas (3,2) can be induced if P has moving power, which is not so plausible.

What did happen in the Pursuit Game? As the Egyptians bore down on the Israelites, Moses responded to their cries:

Have no fear! Stand by, and witness the deliverance which the LORD will work for you today; the Egyptians whom you see today you will never see again. The LORD will battle for you; you hold your peace! (Exod. 14:13–14)

Indeed, with the help of God, Moses parted the Red Sea before the Israelites and, as soon as they had safely crossed it, directed that the water flow back. The entire Egyptian army was swallowed up in its waters.

In retrospect, Pharaoh's tenacious pursuit of the Israelites was for him an unmitigated disaster, but I would argue that he could not possibly have anticipated this outcome when he made his strategy choice. At worst, Pharaoh might have anticipated something on the order of the plagues, each of which at least provided a warning of worse things to come. It would have been riskier, I assume, for Pharaoh to have suffered the loss of the slaves without a fight, which might have led the Egyptians to rise up in anger against him and perhaps depose or kill him.

By contrast, prospects of recapturing the Israelites looked auspicious. In fact, as the Egyptians closed in on the petrified Israelites, the Israelites were on the verge of deserting their leader:

Greatly frightened, the Israelites cried out to the LORD. And they said to Moses, "Was it for want of graves in Egypt that you brought us to die in the wilderness? What have you done to us, taking us out of Egypt? Is this not the very thing we told you in Egypt, saying, 'Let us be, and we will serve the Egyptians, for it is better for us to serve the Egyptians than to die in the wilderness?'" (Exod. 14:10–12)

Furthermore, even after God threw the Egyptians into a panic by locking their chariot wheels as they crossed the parted sea, they were not beyond making a clearheaded calculation that they thought might save them:

Let us flee from the Israelites, for the LORD is fighting for them against Egypt. (Exod. 14:25)

But by then it was too late.

Now counterpose this interpretation of a rational Pharaoh and calculating Egyptians against God's preview of the whole affair:

Now I will stiffen the hearts of the Egyptians so that they go in after them; and I will assert My authority against Pharaoh and all his warriors, his chariots and his horsemen. Let the Egyptians know that I am the LORD, when I assert My authority against Pharaoh, his chariots, and his horsemen. (Exod. 14:17–18)

Without contesting God's own glory-seeking motives, which are completely consistent with His previous behavior, it now must be asked whether a one-person interpretation of the Pursuit Game, with God/ Moses as the sole player, is consistent with the outcome. Specifically, if God had made Pharaoh's strategy choice for him, as in the Defiance (Manipulated) Game, would He have made the choice Pharaoh did in the Pursuit Game?

This question can be answered by deleting P's preferences from the outcome matrix in figure 9.2. Then, assuming her "choices" are really states of nature that SB controls, as was true for Pharaoh in the Defiance (Manipulated) Game in figure 9.1, what would SB do? Obviously, based on his preferences, he would have P choose pursue—in SB's terms, make P obstinate—and then help P's adversary, as actually occurred in the Bible, to provoke a new confrontation that leads to SB's best outcome. In this manner, a manipulative SB, effectively controlling P's thoughts and actions, can also be used to explain what happened in the Pursuit Game.

Thus, we are not able to distinguish whether, in the Bible, Pharaoh was truly his own man, as suggested by the two-person game (figure 9.2), or just a puppet, as suggested by the one-person game (figure 9.1). Both interpretations "work" in the sense of explaining the outcome that occurred in the Bible.

My own predisposition is toward the two-person interpretation, which endows P with free will, since it is consistent with the place reserved for humans in the world that God initially created (see section 1.3). Nevertheless, a case can be made for the puppet interpretation, as represented by the Defiance (Manipulated) Game, for the continuation of the plagues.

Perhaps the truth lies somewhere in between: While Pharaoh was never really his own man, he exhibited greater independence in his later confrontation with God/Moses than his earlier one. His increasing self-assertiveness, I would add, seems undergirded by a shrewd rationality.

9.5   The Salvation Game: Moses and the Israelites at Mount Sinai

Pharaoh was not the only person to have become more assertive. Increasingly, Moses, in his dealings with Pharaoh, shed the apprehensions and fears he had suffered earlier. But it is only after the demise of Pharaoh and the Egyptian army, when Moses faced unexpected internal challenges, that he emerged as a strong, independent figure.

As usual, God set the stage for later trouble by laying down "a fixed rule" (Exod. 15:25) designed to test the Israelites, now wandering in the wilderness:

If you will heed the LORD your God diligently, doing what is upright in His sight, giving ear to His commandments and keeping all His laws, then I will not bring upon you any of the diseases that I brought upon the Egyptians, for I the LORD am your healer. (Exod. 15:26)

Moses does not have long to wait before the Israelites start grumbling about conditions in the wilderness. First, there is not enough bread to eat, which God rectifies by raining down bread from the heaven. Then, when there is no water, God through Moses provides for it. God also helps Joshua, a close associate of Moses, defeat the attacking Amalekites.

Although the Israelites are not always pious, as when some of them fail to observe the Sabbath, there is no serious challenge to God until three months after their departure from Egypt, when they arrive at Mount Sinai:

Now Mount Sinai was all in smoke, for the LORD had come down upon it in fire; the smoke rose like the smoke of a kiln, and the whole mountain trembled violently. The blare of the horn grew louder and louder. As Moses spoke, God answered him in thunder. The LORD came down upon Mount Sinai, on the top of the mountain, and the LORD called Moses to the top of the mountain and Moses went up. (Exod. 19:18–20)

Moses, and later Aaron, are the only people allowed to ascend the mountain. In the miasma enveloping Mount Sinai, God then speaks; His words are the Ten Commandments, the basic laws to which He demands adherence if the covenant is to be consummated and Israel is to become a "holy nation" (Exod. 19:6).

God next relates to Moses, in considerable detail, the rules that must be followed and the punishments for disobedience. Also, Moses is given meticulous instructions on building the Tabernacle, the sacred sanctuary for the worship of God.

Only after Moses ascends the mountain once again, this time to stay for forty days and nights, do the people become restless:

When the people saw that Moses was so long in coming down from the mountain, the people gathered against Aaron and said to him, "Come, make us a god who shall go before us, for that man Moses, who brought us from the land of Egypt—we do not know what happened to him." (Exod. 32:1)

Aaron then became involved in a serious breach of faith, only later to try to retract. He told the people:

"Take off the gold rings that are on the ears of your wives, your sons, and your daughters, and bring them to me." … Then he took them and cast in a mold, and made it into a molten calf. And they exclaimed, "This is your god, O Israel, who brought you out of the land of Egypt!" When Aaron saw this, he built an altar before it, and Aaron announced: "Tomorrow shall be festival of the LORD!" (Exod. 33:2–5)

The revelry of the people at the base of the mountain infuriated God, who commanded Moses:

Hurry down, for your people, whom you brought out of the land of Egypt, have acted basely. They have been quick to turn aside from the way that I enjoined upon on them. (Exod. 32:7–8)

The next exchange between God and Moses firmly establishes Moses as his own man:

The LORD further said to Moses, "I see that this is a stiffnecked people. Now, let Me be, that My anger may blaze forth against them and that I may destroy them, and make of you a great nation." But Moses implored the LORD his God, saying, "Let not Your anger, O Lord, blaze forth against Your people, who You delivered from the land of Egypt with great power and a mighty hand. Let not the Egyptians say, 'It was with evil intent that He delivered them, only to kill them off in the mountains and annihilate them from the face of the earth.'" (Exod. 32:9–12)

Like rhetorical questions a defense attorney might ask, the entreaties of Moses highlight the issue that God, acting as judge, must decide. Given His commitment to Israel, and given also that this commitment is now well known, is it rational for God at this juncture to brush aside His handiwork out of pique?

This is the nub of the matter for God, as Moses brilliantly perceives. On this Moses now builds a cogent defense:

Turn from Your blazing anger, and renounce the plan to punish Your people. Remember Your servants Abraham, Isaac, and Jacob, how You swore to them by Your Self and said to them: I will make your offspring as numerous as the stars of heaven, and I will give to your offspring this whole land of which I spoke, to possess forever. (Exod. 32:12–13)

God, who had earlier recalled in a heavenly metaphor to Moses "how I bore you [children of Israel] on eagles' wings and brought you to Me" (Exod. 19:4), could not turn down Moses's plea, and He relented.

God thus comes off as quick-tempered but merciful. He is, in His own words, an "impassioned god" (Exod. 20:5), but He is also compassionate, perhaps even sentimental. However, it must be stressed—given God's enormous investment in His chosen but "stiffnecked" people, and His strong desire not to destroy His credibility by reneging on His commitment to them—that this "merciful" decision is by no means ill-conceived or fatuous. I will shortly model the game played between Moses and God over the fate of the Israelites, but first I provide more details on the views of the players and the aftermath of their decisions.

It seems that Moses is caught in a bind. After his intercession on behalf of the Israelites, he approached their camp, carrying two tablets on which the Ten Commandments were inscribed. When he "saw the calf and the dancing" (Exod. 32:19), like God, he was enraged:

He hurled the tablets from his hands and shattered them at the foot of the mountain. He took the calf that they had made and burned it; he ground it to powder and strewed it upon the water and made the Israelites drink it. (Exod. 32:19–20)

Moses next wrung a confession out of Aaron for his part in the idolatrous affair. Then, seeing that the people were "out of control" (Exod.

32:25), Moses tried to avert catastrophe by seizing the initiative: "Whoever is for the LORD, come here!" (Exod. 32:26)

Moses's gamble paid off, at least for one tribe in the donnybrook:

And all the Levites rallied to him. He said to them, "Thus says the LORD, the God of Israel: Each of you put sword on thigh, go back and forth from gate to gate through the camp, and slay brother, neighbor, and kin." The Levites did as Moses had bidden; and some three thousand of the people fell that day. And Moses said, "Dedicate yourselves to the LORD this day—for each of you has been against son and brother—that He may bestow a blessing upon you today." (Exod. 32:26–29)

When Moses, who was a Levite, then asked that God pardon the people for their sins, God was forgiving, but not without promising later retribution.

The events I have described in this section offer a rich tapestry of crosscutting cleavages:

• that between God/Moses and the complaining Israelites in the desert;
• that between God and the Israelites, after Moses ascends Mount Sinai, over the building of the molten calf;
• that between Moses and God over the disposition of the Israelites for their idolatry;
• that between Moses and the Israelites (including Aaron), causing Moses to destroy the Ten Commandments and incite the murder of non-Levite Israelites.

The most unusual conflict is between Moses and God over the fate of the Israelites; all the other conflicts involve parties that have behaved badly or been disrespectful to God, which is also true of most of the conflicts analyzed in earlier games.

But the Salvation Game between P and SB, depicted in figure 9.3, is a different matter, because P and SB (i.e., Moses and God) in the Bible story have heretofore been on the same side, and they will remain so. But in the Salvation Game, which is game 21 in the appendix, they differ on how to deal with the Israelites who have sinned, wherein P's choices are to plea (P) or not plea ($\overline{P}$) for their salvation, and SB's choices are to renege (R) or not renege ($\overline{R}$) on his threat to "destroy them" (Exod. 32:9) for their idolatry.

I assume that the primary and secondary goals of the players are as follows:

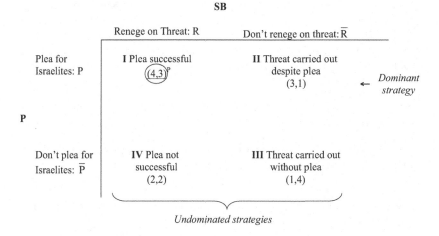

Key: (x,y) = (payoff to SB, payoff to P)
   4 = best; 3 = next-best; 2 = next-worst; 1 = worst
   Nash equilibrium underscored
   Nonmyopic equilibrium (NME) circled
   Superscript P indicates outcome P can induce with a compellent threat

Figure 9.3
Outcome and payoff matrix of the Salvation Game (game 21)

SB: (i) wants to renege if P pleas for the Israelites' salvation but otherwise not to renege; (ii) prefers that P not plea for the Israelites' salvation.

P: (i) wants to plea for the Israelites' salvation; (ii) prefers that SB renege.

The unique Nash equilibrium and NME, (4,3), is that the plea is successful, as was the case. But in the biblical story, it was not Moses's compellent threat—by committing to P, his dominant strategy—that solidified this outcome but rather his persuasive arguments that induced God to choose R, especially after Moses pointed out the costs of wiping the slate clean and having to start anew to build a great nation.

As we have seen, however, there were severe repercussions for non-Levite Israelites after their idolatrous behavior. Nonetheless, this does not end their wayward behavior. Striking figure that Moses now cuts, he cannot, even with God's support, impose a stable order on, or evoke continuing contrition from, his people. His task, perhaps, is not always helped by a still angry God, who instructs Moses to tell the Israelites:

You are a stiffnecked people. If I were to go in your midst for one moment, I would destroy you. Now, then, leave on your finery, and I will consider what to do to you. (Exod. 33:5)

God later followed up this implicit threat with a more magnanimous statement:

I will make all My goodness pass before you, and I will proclaim before you the name LORD, and the grace that I grant and the compassion that I show. (Exod. 33:19)

However, He showed how deep-seated His animosity ran when He said that He was one who "visits the iniquity of fathers upon children and children's children, upon the third and fourth generations" (Exod. 34:7). Thus God, while someone of "grace" and "compassion," seems more ready to rely on imprecations and threats than promises.

Despite God's wariness and cynicism, He renewed the covenant with the Israelites, and Moses delivered the Ten Commandments intact. The Israelites were asked to make sacrifices and bring offerings to God, and eventually the Tabernacle, or sanctuary for the divine, was completed.

The wanderings of the Israelites in the wilderness are next picked up in the Book of Numbers, where, characteristically, the Bible reports:

The people look to complaining bitterly before the LORD. The LORD heard and was incensed: a fire of the LORD broke out against them, ravaging the outskirts of the camp. The people cried out to Moses. Moses prayed to the LORD, and the fire died down. (Num. 11:1–2)

Further problems crop up, God's rancor grows, and a troubled Moses asks:

Why have You dealt ill with Your servant and why have I not enjoyed Your favor, that You have laid the burden of all this people upon me? Did I conceive this people, did I bear them, that You should say to me, "Carry them in your bosom as a nurse carried an infant," to the land that You have promised on oath to their fathers? (Num. 11:11–12)

As Cain rejected fraternal responsibility (see section 8.2), Moses now abjures, metaphorically, maternal responsibility. Moses, however, is no intemperate Cain ready to exact revenge by wreaking violence on the community. Rather, his lugubrious solution is to withdraw:

I cannot carry all this people by myself, for it is too much for me. If You would deal thus with me, kill me rather, I beg You, and let me see no more of my wretchedness! (Num. 11:14–15)

Moses's threat of withdrawal convinces God that His pious but tired servant needs help, and He orders Moses to assemble seventy elders to share with him the burden of governance. God also squelches further grumbling against Moses, including some from Aaron and Moses's sister, Miriam.

I believe Moses and God's interactions after the idolatry at Mount Sinai can be viewed as a reenactment of the Salvation Game (figure 9.3), except Moses as P is pleading not for the Israelites as much as for himself—that he be relieved of caring alone for his insubordinate people. God, as SB, is not reneging on a threat as much as backing away from letting the burden for dealing with the fractious Israelites rest entirely on Moses's shoulders.

Next, the land of Canaan, flowing "with milk and honey" (Num. 13:27), is explored. The people clamor to occupy the land, but it is inhabited by men of great size living in large and fortified cities. Once again the people despair of Moses's and Aaron's leadership:

"If only we had died in the land of Egypt," the whole community shouted at them, "or if only we might die in the wilderness! Why is the LORD taking us to that land to fall by the sword? Our wives and children will be carried off! It would be better for us to go back to Egypt!" And they said to one another, "Let us head back for Egypt." (Num. 14:2–4)

But it is God who is most disturbed that the people "spurn" Him (Num. 14:11), and it is Moses once again who intercedes on their behalf, repeating to God his argument that God's credibility would be impaired if He retracted His oath to His chosen people. This time, though, God, while accepting Moses's plea to spare the Israelites—in another incarnation of the Salvation Game—is not ambiguous about the punishment He intends:

None of the men who have seen My Presence and the signs that I have performed in Egypt and in the wilderness, and who have tried Me these many times and have disobeyed Me, shall see the land that I promised on oath to their fathers. (Num. 14:22–23)

More specifically,

Not one shall enter the land in which I swore to settle you—save Caleb son of Jephunneh and Joshua son of Nun. Your children who, you said, would be carried off—these will I allow to enter; they shall know the land you have rejected. But your carcasses shall drop in this wilderness, while your children roam the wilderness for forty years, suffering for your faithlessness, until the last of your carcasses is down in the wilderness. You shall bear your punishment for forty years, corresponding to the number of days—forty days—that you scouted the land: a year for each day. (Num. 14:30–34)

Fitting retribution, it seems, to ensure that "you shall know what it means to thwart Me" (Num. 14:34).

This punishment does not end troubles for Moses or God. Elements of the Israelites continue to voice complaints—even challenge—Moses's authority again. They are summarily dispatched by a brooding God. As for Moses and Aaron, because God considers them to have been disloyal, "for, in the wilderness of Zin, when the community was contentious, you disobeyed My command" (Num. 27:14), they, too, are prevented from entering the promised land.

Of all the men twenty years or over who originally had fled Egypt, only Caleb and Joshua, who "remained loyal to the LORD" (Num. 32:12), are allowed to cross the Jordan River and set foot in Canaan. Indeed, it is Joshua, with Moses's blessing, who leads the Israelites into the land that God had promised their forefathers would be theirs.

Perhaps the central puzzle in the wanderings of the Israelites in the wilderness is how they were able to survive at all. Almost from the start, God was continually out of sorts and frustrated by their misdeeds, ready on more than one occasion to wipe them out altogether. Their gross misbehavior at Mount Sinai, with the complicity of Aaron, occurred only three months after their departure from Egypt and involved a direct challenge to the first two of the Ten Commandments:

You shall have no other gods before Me.
You shall not make for yourself a sculptured image, or any likeness or what is in the heaven above, or on the earth below, or in the waters under the earth. (Exod. 20:3–4)

My explanation for God's tolerance of the repeated transgressions of the Israelites is that it had little to do with any great compassion He felt for them. Although they were His chosen people, and although He had made an oath to their forefathers to protect them, God found the progeny stubborn and generally disagreeable, if not downright wicked.

How does one deal with such an unfortunate state of affairs? In God's case, He looked for His solution in the best of the lot—Moses— with the hope of setting him up as the protector of the Israelites.

As I showed, however, God had to work hard, especially in the beginning, to convince a reluctant Moses that he was up to the task. But in the end, with his mettle tested and established in his struggle with Pharaoh, Moses was ready and able to serve as a buffer between God and His chosen, if ungrateful and unregenerate, people.

Thus, I interpret Moses's primary role to be a foil for God's anger and explosive temper. God, in my opinion, understood that His patience would be tried to the limit by the querulous Israelites, and He wanted to provide them with a worthy protector to counter any precipitous actions He might take or even contemplate. Moses—until almost the end—served admirably in this role, interceding with God on several occasions to deflect His slashing anger and jealousy. Usually he would make his appeal by playing on God's vanity, continually reminding Him of how much would be lost in future credibility if He reneged on His oath and exterminated the Israelites.

But God could not morosely sit by and do nothing when challenged by egregious breaches of faith. Accordingly, He did punish the apostates after defaming them, yet always—with prodding from a now charismatic Moses—was persuaded to stop short of wiping the slate clean.

Often God tried to root out just the bad elements. But because these were ubiquitous, the job was never quite done. Even Moses, for all his integrity, did not escape unblemished in God's eyes, but that is probably an occupational hazard of a public defender who takes the side of—or at least is associated with—the criminal element in the face of such an imposing authority as God's. Nevertheless, as reported in Deuteronomy, Moses survived his difficult role with unique stature:

Never again did there arise in Israel a prophet like Moses—whom the LORD singled out, face to face, for the various signs and portents that the LORD sent him to display in the land of Egypt, against Pharaoh and all his courtiers and his whole country, and for all the great might and awesome power that Moses displayed before all Israel. (Deut. 34:10–12)

I believe the puzzle of a frustrated and impatient God is solved once one understands that God set up Moses to deal with intractable situations that He knew He would not be able to handle Himself. In particular, recognizing His own quick temper, God deliberately selected a man capable of defusing it. Thereby Moses acted not only as a kind of savior of his people from an all-but-certain debacle in Egypt, but, just as important, also as a guide of the Israelites in the wilderness for forty years. Although Moses was denied the reward of entering Canaan, he was "singled out, face to face" (Deut. 34:10) before

God—which some have taken to mean that Moses was the only mortal ever to see God.[3]

In summary, I see all the major players in the Salvation Game, which is really a set of games, as acting rationally:

• God, in backing off from His outrage at the misbehavior of the Israelites and allowing Moses to intercede on their behalf, bringing out God's merciful side without, at the same time, undermining His future credibility;
• Moses, in accepting his role as the Israelite leader, albeit reluctantly at times, and finally receiving the unique reward of meeting God, but without entering the promised land;
• The Israelites, in rebelling when conditions were difficult, but sometimes having them alleviated and always being saved from total destruction.

While God and Moses are indistinguishable when they act in concert, is God really superior to Moses—in what each does for the Israelites—when they choose different paths? For the most part, they need each other, but God also needs antagonists to succeed in cementing His reputation as a formidable adversary who must be reckoned with for those contemplating resisting Him.

9.6   Conclusions

If there is one character in the Bible who stands apart from other human beings as a near-equal to God, it is Moses. Abraham, although the patriarch and founder of Israel, engaged in morally questionable behavior in offering Isaac as his sacrifice (see section 7.2); Job, while holding steadfast against horrible punishment, still expressed grave doubts about God, who had sanctioned such punishment (see section 7.4). On the other hand, no act of Moses can be challenged as less than righteous, including his killing of the Egyptian and his unflagging efforts to save Israel, even when God did not want to spare the nation and its people.

Moses lived to be 120 years old, and from birth he was blessed. This is not to say that God took direct control of his early life, but the fact that Moses emerged relatively unscathed from a series of taxing

---

3. This interpretation, however, is contradicted by an earlier statement God made to Moses: "But," He said, "you cannot see My face, for man may not see Me and live" (Exod. 33:20).

encounters, including murder, indicates that fate—or God's guiding hand—supported him, especially when he was a timid soul in their first encounters.

Perhaps God calculated that too threatening a figure might encourage Pharaoh to give up his slaves without a fight. In any event, the plodding Moses, with God's coaching and Aaron's forensic skills, did succeed after some early bungling in sustaining a rather grim game against Pharaoh, in which God's power was broadcast to the world when plague after plague was visited upon the hapless but intransigent Egyptians.

That God manufactured the plagues to have their desired effect seems clear; what is less clear is that He later forced Pharaoh, in their next encounter, to pursue the Israelites by making him obstinate once again. What seems closer to the truth, as I argued earlier, is that Pharaoh calculated that he had a good chance of intercepting the slaves while in flight.

If God intervened again on the side of the Israelites, Pharaoh probably did not anticipate his own destruction, at least in one fell swoop. It therefore seems that while God largely established the rules of the game, Pharaoh did not act irrationally within the rules as he perceived them.

Moses, now poised and politically adept, does his best to control an unruly and ungrateful people. In time, though, he becomes disillusioned with, and bitter about, his middleman role and wants out. But just as Moses arrested God's intemperate predilections by courageously pleading for the salvation of the Israelites, God reciprocates by giving His beleaguered disciple help that lightens his burden.

The most striking fact of this final phase of the conflict is not the alliance between God and Moses. This has existed throughout their long relationship, in which Moses emerges as his own man. More surprising is the transformation of a daunting Moses into a strong and forceful personality, which is, I believe, exactly what God intended. It makes Moses not just an acolyte but an independent and, consequently, more efficacious champion of his people, about whom God has gotten very upset, even if they provide Him with an excuse for flexing His muscles to underscore His hegemony.

If Moses succeeds in saving his people in the end, God once again is still able to present magnificent displays of His might and power to keep the people generally in line. These exhibitions, in my opinion, He both needs and wants—and that is the fundamental similarity that this final phase of the conflict has with the preceding phase of confronting Pharaoh. The cast of character changes, but not God's aims, and the stern means He uses to achieve them. Thus, there is perhaps more constancy in the protracted conflict between, at various times, God, Moses, Pharaoh, and the Israelites than first meets the eye.

If there is a new twist in the second phase of the conflict when Pharaoh pursues the Israelites, it comes in Moses's metamorphosis from inept servant to self-assured and adroit spokesman for His people, which is in God's interest but a change He cannot quite live with comfortably. Hence God, while recognizing Moses's rectitude and supreme achievement, denies him a final resting place in Canaan.

Although God's presence in the Defiance (Manipulated) Game, the Pursuit Game, and the Salvation Game is always front and center, His behavior is not always admirable. To me, Moses is the more sympathetic figure and, in his own way, stands out as much as God not only for his beliefs and strategic acumen but also for his ethics (the killing of 3,000 Israelites after their idolatry may be an exception, but it was probably necessary to save the nation from being completely annihilated by God in His rage).

In conclusion, while God and Moses form a team that makes them essentially indistinguishable in the first two games, they diverge in the Salvation Game. In this game, Moses more than holds his own, putting him on a par with God as a strategic thinker.

Indeed, Moses may more than rival God in his moral judgments. What, then, makes SB superior?

10   The Wisdom and Truth Games

10.1   Introduction

All the games I have analyzed so far involve SB (or God) as a player, who is in conflict with a human player. But in this chapter I switch gears.[1] In the first game I analyze, called the Wisdom Game, God appears in the background—possibly standing behind a human player (King Solomon)—just as God in the Defiance (Manipulated) Game and the Pursuit Game (see sections 9.3 and 9.4) stood behind Moses in his battle first with Pharaoh, and then with his fellow Israelites after their idolatry at Mount Sinai.

In these two games, Moses served as God's spokesman, or a surrogate when God did not respond directly to events. However, Moses's partnership with God frayed after the Israelite insurrection at Mount Sinai, when he crossed swords with God to try to save his own tribe, the Levites, from slaughter. In seizing the mantle of leadership in the Salvation Game (see section 9.5), Moses did not so much cross God as try to persuade Him that saving at least some Israelites would be a prudent choice. Moses was successful, and he and God acted in concert once again.

In the Wisdom Game, King Solomon displays a preternatural wisdom in resolving a dispute between two mothers about the maternity of a baby. Outrageously, Solomon proposes to cut the baby in two, giving one half to each mother if they disagree on who the mother is. Fortunately, this does not happen, but the resolution that Solomon engineers is fragile, incapable of being used more than once. It works because the two mothers are unsuspecting of Solomon's (or God's) clever calculation that lies behind it, which the Wisdom Game brings out.

---

1. This chapter is adapted from Brams 2018.

It is, of course, difficult to separate Solomon's wisdom from any coaching that God might have given him, which is never mentioned in the Bible. Is Solomon simply wiser than any of his subjects? If so, then he would appear indistinguishable from a godlike character. Or is he on a par with God—like SB in Newcomb's problem—and able to predict with uncanny accuracy who the true mother is, based on their responses to his edict?

I asked earlier whether SB might be mistaken for a human character; in this chapter, I ask whether Solomon might be mistaken for SB, given the apparent wisdom he shows in finding a solution to the two mothers' dispute. In either case, the question posed is the same: Is SB undecidable—inherently unknowable—or not?

Solomon's decision in the Bible would not be so surprising if he had taken a more cautious approach, not risking the life of the disputed baby. If the rules for handling disputes of this kind were in place, then a more routine resolution, without the drama that Solomon's decision created, would seem not only more attainable but also less dangerous to implement.

In fact, in the second game in this chapter, the Truth Game, I suggest rules that would reveal the truth when two parties make the same claim for something or someone, such as the baby in the Solomon story. These rules, while specific to the dispute the mothers had, can be generalized to other situations.

I believe they provide a sturdier foundation for dispute resolution by giving even sophisticated players, who might otherwise not tell the truth, an incentive to be honest. These rules render the Truth Game, in contrast to the Wisdom Game, invulnerable to manipulation—at least if the players know each other's preferences—inducing the two mothers in the Solomon story to resolve their dispute without ever harming the baby.

10.2 The Wisdom Game: King Solomon's Edict about the Disputed Baby

Most of the "wisdom" of the Bible is simply asserted, as in the Book of Proverbs, which is filled with advice about proper behavior, admonitions against improper behavior, and miscellaneous sayings and aphorisms meant to be instructive on various matters. Lessons, of course, are also meant to be learned from the stories of conflict and intrigue I have analyzed, but the messages in these stories are usually less direct and more often subject to different interpretations.

It is a rare story, indeed, that imbues a character other than God—or one with God at his or her side—with a soaring intelligence and depth of insight that seem to surpass human bounds. True, most characters act rationally according to their goals, and a few like Cain (see section 8.2), and Moses in his later years (see section 9.5), show by the arguments they present to God that they are superb strategists. It is hard, however, to find human characters who, when pitted against fellow mortals, emerge as larger-than-life figures by virtue of their godlike wisdom.

The biblical character who stands out as the most striking exception to this statement is Solomon, who ruled as king of Israel after David, who himself followed Saul, the first king of Israel (see section 8.4). What is usually considered Solomon's most breathtaking judgment is described in just twelve verses in chapter 3 of the First Book of Kings.

This judgment concerns the disposition of a baby boy for whom two women claimed maternity. I begin by modeling this judgment as a game Solomon devised to test the veracity of the two women's claims. Although the game as played involved one woman's moving first, Solomon could as well have set the rules to require simultaneous—or independent, as I assume here—moves and still have achieved the same result.

Solomon's game, the Wisdom Game, arises from the following dispute between two prostitutes who come before him:

The first woman said, "Please, my lord! This woman and I live in the same house; and I gave birth to a child while she was in the house. On the third day after I was delivered, this woman also gave birth to a child. We were alone; there was no one else with us in the house, just the two of us in the house. During the night this woman's child died, because she lay on it. She arose in the night and took my son from my side while your maidservant was asleep, and laid him in her bosom; and she laid her dead son in my bosom. When I arose in the morning to nurse my son, there he was, dead; but when I looked at him closely in the morning, it was not the son I had borne." (1 Kgs. 3:17–21)

The other prostitute protested this version of their encounter: "No, the live one is my son, and the dead one is yours!" (1 Kgs. 3:22). The two women continued arguing in Solomon's presence, while he reflected:

"One says, 'This is my child, the live one, and the dead one is yours'; and the other says, 'No, the dead boy is yours, mine is the live one.'" So the king gave the order, "Fetch me a sword." (1 Kgs. 3:23–24)

Solomon's solution, more cerebral than emotional, was of dazzling simplicity: "Cut the live child in two, and give half to one and half to the other." (1 Kgs. 3:25). The subtlety underlying this solution soon became apparent in the reactions of the two claimants:

But the woman whose son was the live one pleaded with the king, for she was overcome with compassion for her son. "Please, my lord," she cried, "give her the live child; only don't kill it!" The other insisted, "It shall be neither yours nor mine; cut it in two!" (1 Kgs. 3:26)

Then Solomon pronounced judgment:

"Give the live child to her [the first woman]," he said, "and do not put it to death; she is its mother." (1 Kgs. 3:27)

The story concludes with the following observation:

When all Israel heard the decision that the king had rendered, they stood in awe of the king; for they saw that he possessed divine wisdom to execute justice. (1 Kgs. 3:28)

Thus is Solomon venerated for his exemplary judgment, which in the Bible takes on the aura of being "divine."

I assume the primary and secondary goals of the two women are as follows (there is no SB in this story, just the two Ps, though their game is precipitated by a king of divine wisdom, whom I assume to be SB):

Mother: (i) wants to protest Solomon's order; (ii) prefers that the impostor also protest Solomon's order.

Impostor: (i) wants to support Solomon's order; (ii) prefers that the mother protest Solomon's order.

The rationale of the mother is to save her baby by protesting, even if the impostor should gain possession of him. For the impostor, on the other hand, her rationale is to curry Solomon's favor by acceding to his judgment.

The two players' goals give rise to the Wisdom Game shown in figure 10.1 (game 2 in the appendix), wherein I assume the baby would surely be saved if both women protest Solomon's order; this is best (4) for the mother and next worst (2) for the impostor. Next best for the mother (3) is that she protests, and the impostor does not, so the baby

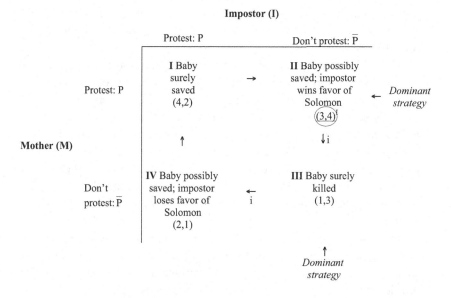

Key: (x, y) = (payoff to M, payoff to I)
  4 = best; 3 = next-best; 2 = next-worst; 1 = worst
  i = impediment
  Nash equilibrium underscored
  Nonmyopic equilibrium (NME) circled
  Arrows indicate progression of states in a weakly cyclic game
  Superscript I indicates outcome that I can induce with either moving power or
    compellent threat power

Figure 10.1
Outcome and payoff matrix of the Wisdom Game (game 2)

is possibly saved, but this is not certain. For the impostor, who does not care about the fate of the baby, she most desires to distinguish herself from the mother, which she can best do (4) when she does not protest and the mother does.

The worst outcomes for the mother (2 and 1) occur when she does not protest; the baby is possibly saved if the impostor alone protests, which is the impostor's worst outcome (1), because she believes that this will cost her Solomon's favor for not supporting his decision. But this is also a poor outcome (2) for the mother, because although the baby might be saved, the mother would be abject for rejecting her baby when the impostor does not.

If neither woman protests, they will not distinguish themselves, and the baby will surely be killed. This is the worst outcome (1) for the mother and the next-best outcome (3) for the impostor, who at least supports Solomon's decision even if she does not distinguish herself from the mother.

In the Wisdom Game, both players have dominant strategies—for the mother to protest, and for the impostor not to do so—which leads to (3,4), the unique Nash equilibrium and NME, whereby the baby is possibly saved. The Wisdom Game is weakly cyclic, with two impediments, in a clockwise direction. Moving power is irrelevant, because both players prefer (3,4), which the impostor can induce, to (1,3), which the mother can induce. If the impostor has threat power, she can reinforce the choice of (3,4) by not protesting, compelling the mother to protest, but this is hardly necessary since the mother has a dominant strategy of protesting.

Fortunately for Solomon, he had foreseen the women's preferences in the Wisdom Game and correctly anticipated that the mother, in particular, would play the game as I have modeled it: Her highest priority would be to save her baby, even at the price of losing him to the impostor. However, Solomon read the women's strategies differently—as evidence of who was telling the truth, which in the end was what he was interested in uncovering.

Solomon's wisdom lay in his setting up a game to distinguish the truthful from the untruthful disputant, which is a form of "mechanism design" that I will say something more about in section 10.3. Here it involves the design of the rules of the game such that play of the game reveals which player is the deceiver (assuming one disputant's claim is truthful and the other's is not). These rules distinguish honest players from dishonest players by eliciting responses that, when properly interpreted, indicate who is lying and who is being truthful.

It is difficult to define "properly interpreted," but one necessary condition is that the mothers not know the game designer's interpretation of their strategy choices. If they did, then presumably they would play a different game from that which the designer intends. Thereby the designer would not elicit the desired truth-revealing responses.

For example, assume that the impostor knew that Solomon did not desire her affirmation of his order but instead intended to favor the woman (women) who protested his order. Then it would obviously be in her interest also to protest, and the game would not distinguish her from the mother.

The designer does, of course, want the disputants to play a game, but the structure of their preferences should not be such that one player has to anticipate the other's choice in order to make a rational choice for herself. That point is illustrated in the Wisdom Game by noting that because each woman has a dominant strategy, it is unnecessary for either to try to predict the other's choice: Whatever the other's choice, each woman's (dominant) strategy is best against it.

To carry this kind of analysis one step further, consider a hypothetical game in which the impostor's preferences are the same as the mother's: Both most prefer a double protest, yielding (4,4), and both least prefer no protest, yielding (1,1). Each would next-most prefer to protest (3) when the other does not (2). In this new game (not shown), the impostor no longer has a dominant strategy of agreeing with the king; instead she, like the mother, has a dominant strategy of protesting, thus assuring the mutually best outcome of (4,4).

This game, however, is not one involving deception but rather one in which information about maternity is fugitive. Naturally, if both women have maternal preferences, and each protests the order, it would not make things easy for Solomon. But well it should not, especially if each woman truly believes she is the mother, and the maternity of the baby cannot be determined from any visual evidence.[2] No game to ferret out the truth can be constructed, even by a Solomon, if the truth is not there to be ferreted out. But can a game be designed in which the truth will come out when there is no hard evidence to determine maternity?

10.3 The Truth Game: A Better Way to Elicit the Truth

Instead of proposing that the baby be cut in two when the dispute between the two mothers arises, suppose that Solomon had proposed the following mechanism for resolving their dispute:

1. Each woman can say the baby is either Mine (M) or Hers (H). If the mothers utter the same strategy (both M or both H), one must be lying, in which case Solomon says that the baby will be placed in an orphanage and both mothers will have to pay for the baby's care there.

2. Today, of course, evidence of maternity would be available from DNA samples or other biological testing, but this was not the case in biblical times. In these times, it would be possible that each mother believes that the live baby is hers if, for example, each gives birth at about the same time, and the babies are then placed next to each other and, along with the mothers, fall asleep. When the mothers awake, and only one baby is alive, each mother might sincerely believe that the live baby is hers if the babies are of the same sex and sufficiently resemble each other.

2. If the women utter different strategies (one says M and the other says H), the baby will go to the mother who utters M, in which case there are two possibilities: Either the mother will get the baby, or the impostor will.

Notice that these rules do not involve anything as harsh as killing the baby. If the women disagree on who the mother is, the baby will be placed in an orphanage.

What gives this mechanism bite is that both women will have to share in the cost of the orphanage, which I assume the mother would willingly do but the impostor would not. More specifically, I assume the primary and secondary goals of the women, in what I call the Truth Game (see figure 10.2), are:

Mother: (i) wants to claim maternity of the baby by saying M; (ii) prefers that the impostor utter H (i.e., say that she is not the mother).

Impostor: (i) does not want the baby to go into the orphanage (off-diagonal outcomes); (ii) prefers that the mother utter H (i.e., say that she is not the mother).

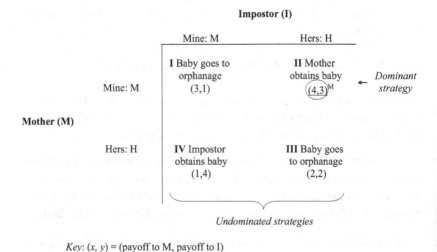

Key: (x, y) = (payoff to M, payoff to I)
  4 = best; 3 = next-best; 2 = next-worst; 1 = worst
  Nash equilibrium underscored
  Nonmyopic equilibrium (NME) circled
  Superscript M indicates outcome that M can induce with compellent threat power

Figure 10.2
Outcome and payoff matrix of the Truth Game (game 14)

The mother's goals need little explanation—she most wants her baby (4), which she will get if the impostor says H, but she won't be devastated if he is placed in the orphanage and she must pay (3). For the impostor, her worst outcomes (2 and 1) are that the baby is placed in the orphanage, because she will have to chip in for his care there. She would be happiest to have the baby herself (4), but having it go to the mother would be a next-best outcome (3), because at least she would not have to pay anything.

Unlike the Wisdom Game, only one player (the mother) has a dominant strategy in the Truth Game, which is game 14 in the appendix. Like the Wisdom Game, however, it leads to a unique Nash equilibrium and NME, (4,3), which gives the mother rather than the impostor (see figure 5.1) her best outcome. If the mother has threat power, she can reinforce this outcome with her compellent threat of saying M.

In the Truth Game, it is the default option of the orphanage, when the women disagree by both saying M or both saying H, that favors the mother. In the Wisdom Game, the women's disagreement when the mother protests and the impostor does not favors the impostor (at least in the impostor's eyes). As we know, however, because the impostor was deceived, Solomon made a decision that favored the mother, though if the impostor had been aware of Solomon's attempt to deceive her and had protested too, Solomon would have been at a loss to determine who the true mother was.

In my opinion, the Truth Game is a much more civilized way of resolving the dispute between the quarreling mothers than the Wisdom Game, whose resolution is fragile: If the impostor had known what Solomon was up to in proposing to cut the baby in two, she would not have been so quick to agree with his decision. Thus, if this game were repeated with either the same or different women, the impostor would know better than to agree with Solomon.

The Truth Game, on the other hand, is not vulnerable to repetition—the players would have no reason to change their strategies if they knew the outcome of earlier play. Also, in this game, eliciting the truth about who the true mother is does not depend on the threat of a dire outcome but rather appeals to the pecuniary instincts of the impostor, who does not want to pay for the support of a child who is not hers.[3]

---

3. The mechanism embodied in the Truth Game is not the only solution to what has been called Solomon's dilemma. While different theorists have proposed alternative mechanisms for determining the identity of the mother, all involve auctioning off the baby, which seems highly unrealistic in the context of the Bible story (Brams 2018).

The Truth Game is one example of a mechanism—a procedure governed by rules—that is designed to induce players to make choices that lead to a desired end. In the case of the Truth Game, the end is to identify the mother of the live baby. Unlike Solomon's Wisdom Game, its ability to distinguish between mother and imposter does not depend on either deception or an incredible threat. Furthermore, it can be repeated without the players' changing their strategies on the basis of what they learned from prior play.

10.4 Conclusions

Recall that the serpent deceived Adam and Eve into eating the forbidden fruit in the garden of Eden by making false claims (see section 6.3). We never learn what possessed the serpent to act in this way; conceivably, it might not have tried if it had known it would be punished by having to crawl and eat dirt "all the days of your life" (Gen. 3:14). Other examples of deception in the Bible include Rahab's deception of her king by sheltering Israelite spies and facilitating their escape; and the Gibeonites' deception of Joshua by pretending that they journeyed from a distant land when this was not the case (Brams [1980] 2003, chap. 6).

In the latter cases, the ends of the deceivers were advanced by their dissembling. Thus, they were rational, but is it ethical to lie and, if so, when?

More concretely, would it have been ethical for Solomon to carry out his gruesome order to cut the disputed baby in two if neither mother had protested his decision? It is pleasing to know that this did not happen because of Solomon's perspicacity in evaluating the responses of the prostitutes and correctly identifying the woman who protested his decision as the mother, but we cannot be sure that this will always happen.

One lesson we might take from this story is that it is perfectly all right for a king to deceive those whose dispute he is arbitrating if it singles out the player who is lying, because the king's deception is critical to producing a just outcome. But, as I noted earlier, while Solomon's probity and wisdom were universally applauded, one can well imagine arbitration games that elicit only half-truths, or do not place the elicited information in a proper context.[4] Greater might, or superior intelligence in this case, does not always make right.

---

4. Bok (1977) offers a good analysis of such questions. See also Brams 1977 and Brams and Zagare 1977, 1981 for game-theoretic models of deception.

Rule of law is supposed to prevent this, but it is, of course, not fool-proof. Unscrupulous individuals, without the judicious temperament of a Solomon, may succeed in sabotaging agreements or subverting institutions, as seen in several Bible stories. Certain problems are ameliorated by having a good knowledge of, and a healthy respect for, the strategic weaknesses in situations.

Morality is empty without safeguards to enforce it. These safeguards may be either explicit or implicit in the nature of the game played, as was the case in the Wisdom Game, whose players were not sophisticated enough to see through Solomon's motives. I judge schemes like Solomon's dangerous, however, because their assumption of a naiveté on the part of the players may be unwarranted. The Truth Game offers a more robust alternative to eliciting the truth, because it can be repeated, even when the players are knowledgeable and sophisticated, with the same results.

Unlike other biblical stories I have analyzed in this book, God is not a player in the Wisdom Game or the hypothetical Truth Game. The latter game is a counterfactual reconstruction of a game that might have been played, showing how Solomon might have produced the same outcome without threatening to kill the baby. The fact that Solomon was human, and implemented a (divine?) solution that in other stories would have required God, suggests the difficulty of distinguishing human players from God.

To be sure, Solomon was not a player in either the Wisdom Game or the Truth Game, but the Wisdom Game, and hypothetically the Truth Game, are mechanisms Solomon did, or could have, put in place. His wisdom shines through in the same way that God's does in other games, which renders Solomon quite undecidable—is he P or SB?

It is hard to say, even though Solomon's subjects extoll his capabilities in the same way they speak of God in other stories. (While a P, Solomon seems inspired by SB.) If there is a difference between P and SB, it is that Solomon seems mainly interested in promoting justice, whereas God in many stories is angry, jealous, or otherwise upset and, almost always, obsessed with His reputation. Paradoxically, God, whose feelings and emotions are often front and center, seems a much more human character than the distant and impersonal Solomon, wise and impartial as he is.

11   Summary and Conclusions

11.1 Undecidability in Decisions and Games

In the first five chapters, I viewed the relationship between P and SB from two different perspectives:

(i) *decisions* that P may make about SB, in which P is uncertain about SB's existence or concern for P; and
(ii) *games* that P may play with SB, in which both players have preferences for the different outcomes.

I recapitulate the main findings from these chapters in this section, and the next five chapters in section 11.2, closing with some final thoughts.

Chapter 1: Introduction
I introduced basic assumptions and concepts in game theory and TOM. I assume that P can exercise free will—she is not a puppet, controlled by SB—which creates challenges for SB, especially in conflict games in which the players' preferences clash and there is no mutually best outcome.

Chapter 2: Belief Decisions
I considered P's choice, as formulated by Pascal, about whether to believe or not believe in God. Although biblical characters almost never doubted the existence of God, some, such as Pharaoh, questioned whether He had the supernatural powers attributed to Him.

Pascal analyzed, from a decision-theoretic viewpoint, the consequences of P's choice when she takes into account the penalties and rewards associated with each of two states—that God exists or does not exist. This became known as Pascal's wager, although Pascal never suggested that betting actually take place.

From a pragmatic, if not secular, perspective, Pascal concluded that belief in God is justified—even if there is only a small probability that He exists—because of the horrific consequences of not believing when, in fact, He exists. Pascal even recommended that P adhere to rituals that reinforce her belief, especially if she thought that she might waver.

I amended Pascal's wager in the Search Decision to include the possibility that the existence of God, whom I subsequently refer to as SB, might be indeterminate. With the inclusion of this option, P does not have a dominant strategy and, therefore, cannot make the relatively easy choice that Pascal's wager affords her.

In the Search Decision, an agnostic might wish to continue searching—though this could prove costly in terms of time, mental effort, or emotional energy—so it seemed reasonable to postulate as a state of nature that SB's existence could be indeterminate. In the related Concern Decision, wherein P must decide whether to be concerned about the choice she makes, there is a similar problem: Whether P benefits or not depends on whether SB is aware of, and cares about, P. Not knowing the answer to this question, P is left in a quandary.

The P that I assume in these decisions, based on her preferences, is neither a believer nor a skeptic but rather someone eager to find answers, unavailing as they may be. In my opinion, the decision-theoretic framework helps to show why P's choices are difficult and may leave her vexed and anxious.

Chapter 3: Belief Games
The two Belief Games that I analyze in this chapter offer a very different perspective from the Belief Decisions. SB is a player, whose existence and behavior cannot be described as simply a state of nature. Like P, SB makes choices to advance his goals; furthermore, it behooves both P and SB to try to take into account the other player's preferences in deciding on a rational strategy.

In both Belief Games, SB's strategies are either to reveal himself or not to do so, and P's strategies are to believe or not to believe in SB. The players' (i) primary and (ii) secondary goals completely define their payoffs in each game, which are the ranks of the possible outcomes from best (4) to worst (1).

In the two Belief Games, SB's goals remain the same: Primarily, he wants P to believe in his existence; secondarily, he does not want to reveal himself, lest doing so prevent a test of P's faith. P, I assume, is of a scientific bent and seeks evidence—positive or negative—for her choice. In both games, she wishes, as a primary goal, that her belief (or nonbelief) in SB's existence be confirmed by evidence (or lack thereof).

What distinguishes the two Belief Games are the different secondary goals of P. In Belief Game 1, P prefers to believe in SB's existence if he exists, whereas she prefers not to believe if he does not exist in Belief Game 2. Thereby P gives SB the benefit of the doubt in Belief Game 1 but not in Belief Game 2.

In both games, SB has a dominant strategy of not revealing himself, inducing P not to believe, which is the unique Nash equilibrium in both games. In Belief Game 1, however, it is Pareto-inferior (worse for both players) to believe with revelation. In this game, P can induce her best outcome—belief with revelation—if she has moving power, but SB can counter if he has moving power by inducing belief without revelation. In Belief Game 2, there is a compromise (3,3) outcome of belief with revelation, which P can induce with moving power but SB cannot, nor can SB induce his best outcome of (4,1).

Chapter 4: Nonmyopic Equilibria in the Belief Games
Belief with revelation is in an NME in both Belief Games, but P would prefer nonbelief and nonrevelation in Belief Game 2, which produces a second NME in this game. In Belief Game 1, belief without revelation is the second NME, which SB would prefer.

Although belief with revelation is an NME in both Belief Games, it competes with a second NME in each game. Surprisingly, it is P's, not SB's, moving power in each game that reinforces the choice of this outcome, suggesting that P should not give up on SB if he does not appear. However, it may take a new generation of Ps for their belief to be restored, provided that there is evidence that SB has revealed himself in these cyclic games.

This cyclicity is perhaps the most important feature of both Belief Games. It suggests that no outcome is likely to be permanent, making any stable outcome elusive. This instability seems confirmed by several Bible stories, wherein God becomes extremely frustrated by His inability to inculcate belief in His subjects, including His chosen people, the Israelites. Then He usually resorts to some kind of punishment, which often creates havoc but also provides evidence that He exists and so must be taken seriously.

If SB is not able to establish his rule once and for all, it is legitimate to ask if he is superior. Besides moving power, omniscience would seem a good measure of superiority, whose effect I consider in the next chapter.

Chapter 5: Paradoxes of Prediction

I discuss Newcomb's problem, in which SB is assumed to have near omniscience. Knowing that SB can almost surely predict her choice, however, theorists do not agree on what P should do—take one box or both boxes. Because of the conflict between the dominance principle and the expected-utility principle, it may, in fact, be rational for P to make a choice not predicted by SB, which calls into question his omniscience.

This inability is even more stark in Prediction Games 1 and 2, especially the latter in which P desires to assert her free will—by making a choice that SB does not predict—than to agree with SB. This game has no Nash equilibrium in pure strategies, so it cycles without impediments.

There are two NMEs in both Prediction Games, one of which P prefers and the other of which SB prefers. Each player can implement his or her preferred outcome with moving power, rendering this power effective. But it also makes predictions hazardous if each player tries to force stoppage of play at his or her preferred NME.

There is a paradox associated with omniscience in the game of Chicken (game 57) as well as five other games. In Chicken, SB's ability to make perfect predictions and P's awareness of SB's ability lead to an outcome that is best for P and only next worst for SB, making SB's predictive abilities, anomalously, a curse rather than a blessing.

In summary, I show in chapters 1–5 that the rationality of believing in SB is anything but straightforward. Moreover, SB's superior powers do not always help, and may in fact backfire by helping P more than SB, at least in terms of the players' comparative rankings of outcomes. This makes the decidability of SB fraught with difficulty, about which several Bible stories create further murkiness and also raise doubts about SB's moral superiority.

11.2 Undecidability in the Bible

In the second five chapters, I analyzed how God's superiority, or lack thereof, is manifest in several stories, all but one of which occur between God and human players.

Chapter 6: The Constraint and Temptation Games

In the Constraint Game, God may or may not impose constraints on Adam and Eve, who may or may not adhere to them if they are imposed. This game is complicated by the wily serpent in a contemporaneous Temptation Game, who successfully tempts Eve. She in turn inveigles Adam into eating from the forbidden fruit.

We never learn why the serpent wished to tempt Eve, but in the end Eve falls prey to its powers of persuasion. Might the serpent be a Satan-like creature that God finds convenient to make His evil surrogate for testing Adam and Eve?

This hypothesis is rendered plausible if God wants Adam and Eve to disobey Him in order to set an example that violation of His commands carries serious consequences. But the fact that God does not kill Adam and Eve for their sins, as He had threatened to do, suggests that He wanted them to fail in order to be able to promulgate both His power and His mercy in punishing them but sparing their lives.

God's real concern, it seems, was that if Adam and Eve ate with impunity from the tree of knowledge of good and bad, they would become godlike and so too much like Him. As a consequence, they might be able to challenge His dominion over all living things.

But God also makes clear His vulnerability—in particular, that a thin line separates Him from them, based on His possessing a kind of moral knowledge. He does not want to share it with humans, it appears, lest they be able to mimic Him. Ironically, however, God's moral superiority is challenged in later stories by the evil that He sanctions.

Chapter 7: Three Testing Games

The three Testing Games highlight this issue. In Testing Game 1, God asks Abraham to prepare Isaac, his son, for a burnt offering. Abraham is obedient to a fault, so the test is successful: Abraham proves himself willing to do something awful, without a peep of objection, and, consequently, Isaac's sacrifice is arrested by the intervention of an angel.

Does this signify that Abraham had complete faith in God and the superiority of His moral judgment? As I reconstructed the players' choices and preferences in Testing Game 1, Abraham could surmise that God has a dominant strategy of reneging on His command to sacrifice Isaac. Supposing that God will choose it, Abraham can afford *not* to choose his own dominant strategy of disobeying God. Instead, he can heed God's command for a sacrifice, which yields an NME that is not the Nash equilibrium.

But there is a second NME, whereby Abraham obeys God's command and God reneges on His demand that Isaac be sacrificed. However, it occurs only if play starts at this outcome, which one might argue it did when Abraham prepares Isaac for sacrifice.

God can implement this NME if He has moving power. However, its invocation is not necessary if Abraham anticipates that God will renege on His command, based on God's promise that Abraham will found a great nation and have multitudinous descendants.

Obviously, this cannot happen if Abraham kills Isaac. Abraham, therefore, can afford to forsake his dominant strategy of disobedience, knowing that Isaac will almost surely be saved and this will please God.

On the one hand, if Abraham outwitted God in this manner, we might question whether God is really the superior player. On the other hand, it seems plausible that God could anticipate that Abraham would make this kind of calculation, in which case God would know that His testing of Abraham is not genuine. If this is the case, can God be considered morally superior if He knows His test has been adulterated, and the results are bogus?

In Testing Game 2, there are also two NMEs, but unlike Testing Game 1, the one that was chosen by God and Jephthah is also the Nash equilibrium. Jephthah, to his great dismay, realizes that he must sacrifice his daughter because God will hold him to his vow and not stop the sacrifice. He probably also knows that pleading for clemency would be fruitless, because if God gives him a pass, it would set a bad precedent: Solemn vows are not revocable.

Did God know that Jephthah would put himself in this predicament when he vowed to sacrifice the first living creature he saw if he was victorious against the Ammonites? If God knew, should He have warned Jephthah against making a foolish vow?

By not doing so, one might consider God a harsh judge of human frailties. But in defense of God's choice, one might also argue that in teaching Jephthah a brutal lesson, God conveyed to others that there are costs in making a thoughtless vow. I hesitate to condemn or praise God for His action in this case, which raises the question of whether it is ethical to use one person to set an example for others.

In Testing Game 3, Job is a paragon of righteousness at the outset, but God still permits him to be tested by Satan. Although the players' strategies are very different, the structure of this game is the same as that of Testing Games 1 and 2; the NME chosen is the same as that in Testing Game 2 but not Testing Game 1.

This time, it is the afflicted player, Job, who eventually succeeds. However, he suffers mightily before triumphing over Satan, in part because, while faltering, he never loses his faith in God entirely. One can say that God is vindicated, obtaining His best outcome, but Job does not do so well when he must endure the worst depredations of Satan. Although Satan loses his wager with God, he apparently survives to another day to test other human players.

In summary, the three people tested by God or Satan all survive their tests, but their experiences are bittersweet:

- Abraham saves Isaac, but there may well be repercussions in their father-son relationship over Abraham's apparent willingness to sacrifice his son.
- Jephthah fails to save his daughter, but he learns a searing lesson about making a vow that he cannot retract; as a small consolation, his daughter's execution, at her request, gets postponed for a summer.
- Job is rewarded with a new family and great riches, but is that really recompense for being severely tested, and losing everything, in order that God can win a wager?

All these stories put God's ethics under scrutiny, which casts doubt on whether He can be considered on a morally higher plane than those He tests.

Chapter 8: The Incitement, Blame, and Deception Games

In the Incitement Game, God may or may not incite Cain to kill his brother, Abel. God does so by downgrading the significance of Cain's offering, compared to Abel's, by ignoring Cain's and praising Abel's offering.

This causes Cain to strike out in anger against Abel, killing him. True, Cain's offering was inferior to Abel's, but it seems that God did nothing to try to avert the anger He stoked in Cain.

The strategic choice Cain faces is to swallow his pride and let this slight pass, or let his anger burst forth. Because Cain's worst outcome is to do nothing when God praises his brother's sacrifice, he reacts with vengeance, leading to the unique Nash equilibrium and NME in the Incitement Game. This outcome is reinforced by God's moving power.

With Abel dead, Cain may or may not defend himself in the Blame Game. He chooses to defend himself not by denying the murder but instead by implicating God in it for inciting his jealousy. Although punished, Cain shames God into making him a marked man, which prevents Cain's own murder. Thus, Cain's defense in the Blame Games works, up to a point, which is better than trying to bottle up his anger.

The Blame Game is strongly cyclic, so it does not have a Nash equilibrium in pure strategies. It does have an NME, whereby Cain defends himself when God marks him.

This is also the outcome if either player has moving power; it can be reinforced by a compellent threat of Cain. However, the fact that it is the unique NMEs means that farsighted players would be inclined to choose it without the exercise of any kind of power.

The same holds in the Deception Game, in which Saul is persuaded to be king but then is systematically undermined by the prophet Samuel, who receives instructions from God. The NME in this game, which is also a Nash equilibrium, favors Samuel and God when they withhold support from Saul—ostensibly, because Saul's offering, like Cain's, was deemed inferior.

Although compellent threats by either Samuel or Saul lend further support to this outcome, it would almost surely be chosen by ordinary players because of its status as the unique Nash equilibrium and NME. This also holds in the Incitement and Blame Games. Therefore, in none of the three games is God's presence as a superior being necessary to explain their outcomes, underscoring the problem of deciding that SB is a player in these games.

Chapter 9: The Defiance (Manipulated), Pursuit, and Salvation Games

The Defiance (Manipulated) Game is not a game, because there is only one player (SB), who must make a decision in the face of uncertainty about the nature of P. Unlike Pascal's wager (see section 2.1), which also involves only one player who must make a decision (to believe or not believe in SB), SB is the player in the Defiance (Manipulated) Game, and the states of nature describe what type P is (submissive or defiant) instead of whether SB exists or not.

In the Bible story, it is God, with Moses as His spokesman, who can stop or continue the plagues to try to force Pharaoh to let the Israelites depart Egypt. The conflict escalates when God hardens Pharaoh's heart. This action robs Pharaoh of free will: He does not make an independent decision, because God, in effect, chooses the state of nature that yields His own best outcome. If there were not this interference by God, His strategies would be undominated, making His choice of a strategy more difficult.

The Pursuit Game, after God loosens His grip on Pharaoh's mind, produces the Defiance (Manipulated) Game, giving both players free rein on what strategies they may choose. Their preferences for the four outcomes, however, do not single out any as a Nash equilibrium, producing instead a strongly cyclic game in which a new confrontation between Moses/God and Pharaoh is the unique NME. Unfortunately for Pharaoh, his forces pursuing the fleeing Israelites are swallowed up in the Red Sea, and Moses/God are triumphant.

There is no question that the Israelites, with God's help, demonstrate their superiority, but one can certainly question the tactics God invokes—first to provoke the confrontation with a series of plagues that Pharaoh could not stop because of God's mind control, and then to lure Pharaoh's forces into a situation they had no ability to prevent.

But SB does not always get his way. The Salvation Game demonstrates how P (Moses in the Bible) levels the playing field with his rhetoric, making a compelling argument that SB's initial judgment is wrong. As a result, Moses is able to persuade God that it would be foolish of Him to annihilate the Israelites for their idolatry at Mount Sinai: His investment in His chosen people would be lost forever.

Instead, Moses takes the position that some punishment of the Israelites is called for, which takes the form of the Levites killing many non-Levites and renewing their pledge to God. In this way, Moses appeases the fury of God, showing himself to have a more even temperament than God in a crisis. For his intercession, however, Moses is prevented from entering the promised land, demonstrating how God holds grudges against those who, even for good reason, oppose Him.

Chapter 10: The Wisdom and Truth Games
In this penultimate chapter, I do not address the superiority of SB as a player directly. Instead, I analyze the "divine wisdom" of Solomon in resolving a dispute between two mothers over the maternity of a baby born to one of them.

I presume that the divinity of Solomon's wisdom was God-endowed, or at least God-inspired. Solomon was the third king of Israel, following the discredited regime of Saul and the celebrated regime of David, who was renowned for his courage but also for his complicity in two murders (of Uriah and Nabal) as well as adultery with their wives (Bathsheba and Abigail).

In the case of Solomon, I question whether his proposal to cut the disputed baby in two demonstrated his superior wisdom. In the Wisdom Game, it is true that Solomon was able to identify the true mother from her and the impostor's responses to his edict, but his choice depended entirely on the impostor not being able to see through the game he set up between the two women.

If the impostor had been more perceptive, Solomon's vaunted wisdom would have revealed the fragility of his solution. He would not have been able to single out the true mother, making his wisdom appear less than divine.

I present a Truth Game that would be strategy-proof in identifying the true mother; moreover, it can be used again in similar disputes in which the truth is fugitive. By rendering truthful responses from the two women as the unique Nash equilibrium and NME in this game, it is more robust than Solomon's solution. To be sure, this new game assumes an orphanage in which the baby could have been placed—with each mother contributing to his care—but even in biblical times, orphanages almost surely were available, especially in kingdoms as prosperous as Solomon's.

11.3   Coda

I have tried to show throughout this book that SB's superiority, in theory as well as practice—at least as represented by stories in the Hebrew Bible—does not necessarily enable him to realize his preferred outcomes. His supernatural powers, defined game theoretically, may not always bolster or improve his position and may even degrade it, as in the paradox of omniscience.

In practice, the biblical God is frequently thwarted in achieving His aims. Then He becomes angry and frustrated with human players, such as Cain and even Moses, who challenge His choices and sometimes persuade Him to retract them.

Other characters, such as Abraham, follow His orders to the letter to try to obtain preferred outcomes for themselves. In Abraham's case, for example, it is plausible that he anticipated God would renege on His command that he sacrifice Isaac. Surmising that he was only being tested, Abraham could do better adhering to God's command. Finally, there are cases—such as Job's tribulations and suffering, which was the product of God's wager with Satan—that lead one to question the morality of the wager and the rectitude of God's ethics.

If SB is (i) not all-powerful, (ii) his presence in human affairs not always evident, and (iii) the morality of his actions sometimes suspect, then I believe that agnosticism is the appropriate theological stance. To the best of my knowledge, there is no theoretical argument or empirical evidence (from the Bible or, for that matter, science) that would render decidable the existence of a superior being. At least I have not discovered it.

Appendix

There are 78 structurally distinct 2×2 ordinal games in which the two players, each with two strategies, can strictly rank the four states from best to worst.[1] These games are "distinct" in the sense that no interchange of the column player's strategies, the row player's strategies, the players, or any combination of these can transform one game into any other. That is, these games are structurally different with respect to these transformations.

Of the 78 games, 21 are no-conflict games with a mutually best (4,4) state. This state is always both a Nash and a nonmyopic equilibrium (NME) in these games; no kind of power—moving, order, or threat—is needed by either player to implement it as an outcome.

I list here the remaining 57 games, in which the players disagree on a most-preferred state. The numbers used in the appendices of Brams (1994, 2011) are shown for each game.[2] The 57 games are divided into three main categories: (i) those with one NME (31 games), (ii) those with two NMEs (24 games), and (iii) those with three NMEs (2 games).

---

1. For a complete listing of the 78 games, see Rapoport and Guyer 1966; Rapoport, Guyer, and Gordon 1976; and Brams 1977, in which the games are divided into three categories based on their vulnerability to deception. Fraser and Kilgour (1986, 1988) enumerate the 726 2×2 games in which preferences are not necessarily strict (i.e., there may be ties in the ranks). Fishburn and Kilgour (1990) simplify preferences further by making only a binary distinction in preferences; Robinson and Goforth (2005) provide a sophisticated mathematical analysis of 2×2 games, giving what they call a "new periodic table"; and Bruns (2015) offers further distinctions, with an elaborate color-coded table.

2. The moving-power outcomes identified in the 57 conflict games listed in Brams 1982, [1983] 2007 differ somewhat from those given here because of changes I have made in the rules of play of TOM. Threat-power outcomes were previously identified in Brams [1983] 2007 and Brams and Hessel 1984. Order power was first used in Brams 1994, which was previously called "staying power" (Brams [1983] 2007; Brams and Hessel 1983); a different concept of staying power is defined in Kilgour, De, and Hipel 1987.

I have grouped together at the end of the list the 9 games with indeterminate states—7 of which fall in category (ii) and 2 of which fall in category (iii)—in which order power is effective when play starts at an indeterminate state. Indeterminate states have two NMEs in the anticipation game (AG), separated by a slash.

Moving power outcomes that the row and column players can induce are indicated by superscripts in the key that follows. If the outcomes induced by moving power are different, then this power is effective if the player who possesses it can implement a better outcome for itself than if the other player possessed it; otherwise, it is irrelevant or ineffective. The latter two categories are distinguished in Brams 1994 (see chapter 4), which also distinguishes cyclic games that are strongly cyclic (no impediments), moderately cyclic (one impediment), and weakly cyclic (two impediments).

Threat power outcomes are indicated by superscripts in the key that follows, with compellent threat outcomes distinguished from deterrent threat outcomes. Threat power is always effective when the row and column players can induce different outcomes; when they induce the same outcome, threat power is irrelevant.

Certain games are further identified by their descriptive names or by the conflicts modeled by them in the text. These games (and their numbers) are, in the order given in chapters 3–10, Belief Games 1 and 2 (games 48 and 49), Prediction Games 1 and 2 (games 48 and 42), Chicken (game 57), Constraint Game (game 25), Temptation Game (game 5), Testing Games 1, 2, and 3 (game 33), Incitement Game (game 19), Blame Game (game 46), Deception Game (game 17), Pursuit Game (game 42), Salvation Game (game 21), Wisdom Game (game 2), and Truth Game (game 14).

The key to the symbols is as follows:

(x,y) = (payoff to row, payoff to column)
$[x,y]$ = [payoff to row, payoff to column] in the AG
$[w,x]/[y,z]$ = indeterminate state in the AG, where $[w,x]$ is the NME induced if row has order power, $[y,z]$ if column has order power
4 = best; 3 = next best; 2 = next worst; 1 = worst
Nash equilibria in original game and in the AG underscored (except when there is only one NME in the AG)
NMEs in original game circled
m/M = moving-power outcome row/column can induce
t/T = threat-power outcome row/column can induce
c/d = compellent/deterrent threat outcome

(i) 31 Games with One NME

Wisdom

1		2		3		4	
(3,4)^{mMTc}	(4,2)	(3,4)^{mMTc}	(4,2)	(3,4)^{mMTc}	(4,1)	(3,4)^{mMTc}	(4,1)
(2,3)	(1,1)	(1,3)	(2,1)	(2,3)	(1,2)	(1,3)	(2,2)

Temptation

5		6		7		8	
(2,4)^{Tc}	(4,2)	(2,4)^{Tc}	(4,1)	(3,3)	(4,2)	(3,3)	(4,2)
(1,3)^{m}	(3,1)^{M}	(1,3)^{m}	(3,2)^{M}	(2,4)	(1,1)	(1,4)	(2,1)

9		10		11		12	
(3,3)	(4,1)	(2,3)	(4,2)	(2,3)	(4,1)	(3,4)^{mMTc}	(4,1)
(1,4)	(2,2)	(1,4)	(3,1)	(1,4)	(3,2)	(2,2)	(1,3)

Truth

13		14		15		16	
(3,4)^{mMTc}	(4,1)	(3,4)^{Tc}	(2,2)	(3,4)^{Tc}	(2,1)	(3,4)^{Tc}	(1,2)
(1,2)	(2,3)	(1,3)	(4,1)	(1,3)	(4,2)	(2,3)	(4,1)

Deception / Incitement

17		18		19		20	
(3,4)^{Tc}	(1,1)	(2,4)^{Tc}	(3,2)	(2,4)^{Tc}	(3,1)	(3,4)^{Tc}	(2,3)
(2,3)	(4,2)	(1,3)	(4,1)	(1,3)	(4,2)	(1,2)	(4,1)

Salvation

21

$(3,4)^{Tc}$	$(1,3)$
$(2,2)$	$(4,1)$

22

$(2,4)^{Tc}$	$(3,3)^{td}$
$(1,2)$	$(4,1)$

23

$(3,3)^{mMTc}$	$(4,1)$
$(2,2)$	$(1,4)$

24

$(3,3)^{mMTc}$	$(4,1)$
$(1,2)$	$(2,4)$

Constraint

25

$(3,2)^{M}$	$(4,1)$
$(2,3)^{m}$	$(1,4)$

26

$(3,2)^{M}$	$(4,1)$
$(1,3)^{m}$	$(2,4)$

27

$(2,3)$	$(4,1)$
$(1,2)$	$(3,4)^{mMtcTd}$

28

$(2,2)$	$(4,1)$
$(1,3)$	$(3,4)^{mMtcTd}$

29

$(3,2)$	$(2,1)$
$(4,3)^{mMtdTc}$	$(1,4)$

30

$(2,2)$	$(4,1)$
$(3,3)^{mMtdTc}$	$(1,4)$

31

$(2,2)$	$(3,1)$
$(4,3)^{mMtdTc}$	$(1,4)$

(ii) 24 Games with Two NMEs

Prisoners' Dilemma

32

$(2,2)$	$(4,1)$
$[2,2]$	$[3,3]$
$(1,4)$	$(3,3)^{tdTd}$
$[3,3]$	$[\underline{3,3}]$

Testing 1/2/3

33

$(3,4)^{MTc}$	$(4,3)^{mtd}$
$[\underline{3,4}]$	$[4,3]$
$(1,2)$	$(2,1)$
$[3,4]$	$[3,4]$

34

$(3,4)^{MTc}$	$(4,3)^{mtd}$
$[\underline{3,4}]$	$[4,3]$
$(2,2)$	$(1,1)$
$[3,4]$	$[3,4]$

35

$(2,4)^{Tc}$	$(4,3)^{mMtd}$
$[\underline{2,4}]$	$[4,3]$
$(1,2)$	$(3,1)^{M}$
$[2,4]$	$[2,4]$

36

$(3,4)^{MTc}$	$(4,3)^{mtd}$
$[\underline{3,4}]$	$[4,3]$
$(2,1)$	$(1,2)$
$[3,4]$	$[3,4]$

37

$(3,4)^{MTc}$	$(4,3)^{mtd}$
$[\underline{3,4}]$	$[4,3]$
$(1,1)$	$(2,2)$
$[3,4]$	$[3,4]$

38

$(3,4)^{MTc}$	$(4,2)^{m}$
$[\underline{3,4}]$	$[4,2]$
$(2,1)$	$(1,3)$
$[3,4]$	$[3,4]$

39

$(3,4)^{MTc}$	$(4,2)^{m}$
$[\underline{3,4}]$	$[4,2]$
$(1,1)$	$(2,3)$
$[3,4]$	$[3,4]$

Prediction 2
Pursuit

40

(3,3)^MTc	(4,2)^m
[3,3]	[4,2]
(2,1)	(1,4)
[3,3]	[3,3]

41

(3,3)^MTc	(4,2)^m
[3,3]	[4,2]
(1,1)	(2,4)
[3,3]	[3,3]

42

(2,4)^M	(4,1)
[2,4]	[3,2]
(3,2)^m	(1,3)
[3,2]	[2,4]

43

(2,4)^M	(3,1)
[2,4]	[4,2]
(4,2)^m	(1,3)
[4,2]	[2,4]

Blame

44

(2,3)^M	(4,1)
[2,3]	[3,2]
(3,2)^m	(1,4)
[3,2]	[2,3]

45

(2,3)^M	(3,1)
[2,3]	[4,2]
(4,2)^m	(1,4)
[4,2]	[2,3]

46

(3,4)^MtcTd	(2,1)
[3,4]	[4,2]
(4,2)^m	(1,3)
[4,2]	[3,4]

47

(3,3)^MtcTd	(2,1)
[3,3]	[4,2]
(4,2)^m	(1,4)
[4,2]	[3,3]

Belief 1
Prediction 1
48

(2,3)	(4,2)^m
[3,4]	[4,2]
(1,1)	(3,4)^MTd
[3,4]	[3,4]

9 Games with Indeterminate States

Belief 2
49

(2,4)^Tc	(4,1)
[3,3]	[2,4]
(1,2)	(3,3)^Mtc
[2,4]/[3,3]	[2,4]

50

(2,4)^Tc	(4,3)^mMtd
[4,3]	[4,3]
(1,1)	(3,2)
[2,4]/[4,3]	[4,3]/[2,4]

51

(3,4)^Tc	(2,1)
[4,3]	[3,4]/[4,3]
(1,2)	(4,3)^tc
[3,4]/[4,3]	[3,4]

	52		53		54
(2,4)^Tc ⃝	(3,1)	(2,3)	(3,4)^Tc ⃝	(2,2)	(3,4)^Tc ⃝
[4,3]	[4,3]/[2,4]	[3,4]/[4,2]	[4,2]	[3,4]/[4,3]	[4,3]
(1,2)	(4,3)^tc ⃝	(4,2)^tc ⃝	(1,1)	(4,3)^tc ⃝	(1,1)
[2,4]/[4,3]	[2,4]	[3,4]	[3,4]/[4,2]	[3,4]	[3,4]/[4,3]

	55
(2,2)	(4,3)^tc ⃝
[3,4]/[4,3]	[3,4]
(3,4)^Tc ⃝	(1,1)
[3,4]	[3,4]/[4,3]

(iii) 2 Games with Three NMEs

	56		Chicken 57
(2,4)^Tc ⃝	(4,2)^m ⃝	(3,3) ⃝	(2,4)^Tc ⃝
[3,3]	[4,2]	[3,3]	[4,2]/[3,3]
(1,1)	(3,3)^Mtc ⃝	(4,2)^tc ⃝	(1,1)
[2,4]/[3,3]	[2,4]	[3,3]/[2,4]	[2,4]/[4,2]

Glossary

This glossary contains definitions of the more technical terms and concepts used in this book. Illustrations of them can be found in the text.

Anticipation game (AG)   An anticipation game is described by a payoff matrix, whose entries, given in brackets, are the nonmyopic equilibria (NMEs) into which each state of the original game goes.

Backtracking   Backtracking occurs when a player moves first in one direction and then reverses the direction of its move, which is not allowed by the theory of moves (TOM).

Backward induction   Backward induction is a reasoning process in which players, working backward from the last possible move in a game, anticipate each other's choices in order to make a rational choice.

Blockage   Blockage occurs when it is not rational, based on backward induction, for a player to move from a state.

Breakdown outcome/strategy   A breakdown outcome is the Pareto-inferior outcome that a threatener threatens to implement, by choosing its breakdown strategy, unless the threatened player accedes to the threat outcome. See also *Threat outcome/strategy*.

Cardinal utility   See *Utility*.

Chicken   A two-person variable-sum symmetric game in which each player has two strategies: to cooperate or defect. Neither player has a dominant strategy; the compromise outcome, in which both players cooperate, is not a Nash equilibrium, but the two outcomes in which one player cooperates and the other defects are Nash equilibria (game 57 in the appendix).

Common knowledge   Players in a game have common knowledge when they share certain information, know that they share it, know that they know that they share it, and so on ad infinitum.

Compellent threat   In repeated play of a two-person game, a threatener's compellent threat is a threat to stay at a particular strategy to induce the threatened player to choose its (as well as the threatener's) best state associated with that strategy.

Complete information   A game is one of complete information if each player knows the rules of play, the rationality rules, the preferences or payoffs of every player for all possible states, and which player (if either) has moving, order, or threat power. When this is not the case, information is incomplete.

Configuration   See *Game configuration*.

Conflict game   A conflict game is a 2 × 2 ordinal game in which there is no mutually best (4,4) state; a no-conflict game is one in which there is such a state.

Constant-sum (zero-sum) game   A constant-sum (zero-sum) game is a game in which the utility payoffs to the players at every outcome sum to some constant (zero); if the game has two players, what one player gains the other player loses.

Cooperative game theory   In a game, players can make binding and enforceable agreements, usually with respect to how the payoffs will be split among them.

Cyclic game   A 2 × 2 ordinal game is a cyclic game if moves either in a clockwise or a counterclockwise direction never give a player his or her best payoff when the player has the next move. A cyclic game is strongly cyclic (no impediments) if each player always does immediately better by moving; otherwise, it is either moderately cyclic (one impediment) or weakly cyclic (two impediments).

Decidability   See *Undecidability*.

Decision   See *Game against nature (one-person game)*.

Decision theory   Decision theory is a mathematical theory for making optimal choices in situations in which the outcome does not depend on the choices of other players but rather on states of nature that arise by chance.

Deterrent threat   In repeated play of a two-person game, a threatener's deterrent threat is a threat to move to another strategy to induce the threatened player to choose a state, associated with the threatener's initial strategy, that is better for both players than the state threatened.

Dominant strategy   A dominant strategy is a strategy that leads to outcomes at least as good as those of any other strategy in all possible contingencies, and a better outcome in at least one contingency. A *strongly dominant strategy* is a dominant strategy that leads to a better outcome in every contingency, whereas this is not true for a *weakly dominant strategy*.

Dominated strategy   A dominated strategy is a strategy that leads to outcomes no better than those of any other strategy in all possible contingencies, and a worse outcome in at least one contingency. A *strongly dominated strategy* is a dominated strategy that leads to a worse outcome in every contingency, whereas this is not true for a *weakly dominated strategy*.

Effective power   Power is effective when possessing it induces a better outcome for a player in a game than when an opponent possesses it; when the opposite is the case, power is *ineffective*.

Equilibrium   See *Nash equilibrium; Nonmyopic equilibrium (NME)*.

Expected payoff/utility   A player's expected payoff/utility is the weighted sum of the payoff that it receives from each outcome multiplied by the probability of its occurrence.

Extensive form   A game in extensive form is represented by a game tree, in which the players make sequential choices but do not necessarily know all the prior choices of the other players.

Feasibility   A move is feasible if it can plausibly be interpreted as possible in the situation being modeled.

Final state   A final state is the state induced after all rational moves and countermoves (if any) from the initial state have been made, according to the theory of moves (TOM), making it the outcome of the game.

Game   A game is an interdependent decision situation, whose outcome depends on the choices of all players. It is described by rules of play.

Game against nature (one-person game)   A game against nature is a one-person game in which one player is assumed to be "nature," whose choices are neither conscious nor based on rational calculation but instead on chance.

Game configuration   A game configuration is a payoff matrix, in which the initial state is not specified.

Game of conflict   A game of conflict is one in which there is no mutually best outcome.

Game of partial conflict   A game of partial conflict is a *variable-sum game*, in which the players' preferences for outcomes are not diametrically opposed.

Game theory   Game theory is a mathematical theory of rational strategy selection used to analyze optimal choices in interdependent decision situations; the outcome depends on the choices of two or more players, and each player has preferences for the possible outcomes. See also *Cooperative game theory; Noncooperative game theory*.

Game of total conflict   A game of total conflict is a *constant-sum game* in which the players' preferences for outcomes are diametrically opposed: The best outcome for one player is the worst for the other, the next-best outcome for one player is the next-worst for the other, and so on.

Game tree   A game tree is a symbolic tree, based on the rules of play of a game, in which the vertices or nodes of the tree represent choice points, and the branches represent courses of action that can be selected by the players.

Impediment   An impediment in a cyclic game occurs when the player with the next move does immediately worse by moving in the direction of the cycling.

Indeterminate state   A state is indeterminate if the outcome induced from it depends on which player moves first (in which case order power is effective).

Ineffective power   See *Effective power.*

Information   See *Complete information.*

Initial state   An initial state is the state in a payoff matrix where play commences.

Irrelevant power   Moving power is irrelevant when the outcome that can be induced by one player in a game is better for both players than the outcome that can be induced by the other player; threat power is irrelevant when either player's possession of it leads to the same outcome.

Lexicographic decision rule   A lexicographic decision rule enables a player to rank states on the basis of a most important criterion ("primary goal"), then a next-most important criterion ("secondary goal"), and so on.

Mixed strategy   A mixed strategy is a strategy that involves a random selection from two or more pure strategies, according to a particular probability distribution.

Move   In a normal-form game, a move is a player's switch from one strategy to another in the payoff matrix.

Moving power   In a cyclic game, moving power is the ability to continue moving when the other player must eventually stop; the player possessing it uses it to try to induce a preferred outcome.

Nash equilibrium   A Nash equilibrium is a state—or, more properly, the strategies associated with a state—from which no player would have an incentive to depart unilaterally because departure would immediately lead to a worse, or at least not a better, state.

Newcomb's problem   Newcomb's problem describes an apparent conflict between two principles of choice: dominance and expected utility.

Noncooperative game theory   In a game, players cannot make binding or enforceable agreements but can choose strategies and can move to and from states.

Nonmyopic equilibrium (NME)   In a two-person game, a nonmyopic equilibrium is a state from which neither player, anticipating all possible rational moves and counter-moves from the initial state, would have an incentive to depart unilaterally because the departure would eventually lead to a worse, or at least not a better, outcome. *Boomerang NMEs* are states from which the players have an incentive to move from the original state and back to it again.

Normal (strategic) form   A game is represented in normal (strategic) form when it is described by a payoff matrix in which the players independently choose their strategies. The possible states of the game correspond to the cells or entries of the payoff matrix.

Omniscience   In a two-person game, an omniscient player is one who can predict the nonomniscient player's strategy before the nonomniscient player chooses it.

One-person game   See *Game against nature (one-person game)*.

Order power   In a two-person game, order power is the ability of one of the players to dictate the order of moves in which the players depart from an indeterminate initial state in order to ensure a preferred outcome for the player with order power.

Ordinal game   An ordinal game is a game in which each player can order or rank the states but not necessarily assign numerical payoffs or cardinal utilities to them; when there are no ties in the ranking, the ordering is strict.

Outcome   An outcome in the theory of moves (TOM) is the final state of a game, from which no player chooses to move and at which the players receive their payoffs.

Outcome matrix   In a game in normal (strategic) form, the entries of an outcome matrix indicate the outcomes to the players resulting from their strategy choices.

Paradox of omniscience   In a two-person game, a paradox of omniscience occurs when it is in a player's interest to be nonomniscient rather than omniscient.

Pareto-inferior state   A state is Pareto-inferior if there exists another state that is better for all players, or better for at least one player and not worse for any other players.

Pareto-optimal/superior state.   A Pareto-optimal state is one that is not Pareto-inferior. A state is Pareto-superior to a (Pareto-inferior) state if it is better for all players, or better for at least one player and not worse for any other players.

Pascal's wager Pascal's wager is to bet on the existence of God, based on payoffs associated with the choice of believing or not believing in God.

Payoff A payoff is a measure of the value that a player attaches to a state in a game; usually payoffs are taken to be cardinal utilities, but here they are assumed to be ordinal (ranks from best to worst).

Payoff matrix In a game in normal (strategic) form, the entries of a payoff matrix indicate the payoffs to the players resulting from their strategy choices.

Payoff termination Payoff termination occurs when one player receives his or her best payoff in the move-countermove process and, if he or she has the next move, terminates play.

Player See *Rational player.*

Power See *Effective power; Irrelevant power; Moving power; Order power; Threat power.*

Preference Preference is a player's ranking of states from best to worst.

Prisoners' Dilemma A two-person variable-sum symmetric game in which each player has two strategies, cooperate or defect. Defect dominates cooperation for both players, even though the mutual-defection outcome, which is the unique Nash equilibrium in the game, is worse for both players than the mutual-cooperation outcome (game 32 in the appendix).

Pure strategy A pure strategy is a single specific strategy. See also *Mixed strategy.*

Rational choice A rational choice is a choice that leads to a preferred outcome, based on a player's goals.

Rational outcome See *Nash equilibrium; Nonmyopic equilibrium (NME).*

Rational player A rational player is an actor with free will who makes rational choices, in light of the presumed rational choices of other players in a game, or with respect to the states of nature that may arise in a decision.

Rational termination Rational termination is a constraint, assumed in the definition of a nonmyopic equilibrium (NME), that prohibits a player from moving from an initial state unless it leads to a better outcome before cycling.

Rationality rules Rationality rules specify when players will stay at, or move from, states, taking into account the rational moves of the other players.

Rules of play The rules of play of a game describe the possible choices of the players at each stage of a game.

Sequential game A sequential game is one in which players can move and countermove after their initial strategy choices according to the theory of moves (TOM).

State A state is an entry in a payoff matrix from which the players may move. Play of a game starts at an initial state and terminates at a final state, or outcome.

Strategy In a game in normal (strategic) form, a strategy is a complete plan that specifies the course of action a player will follow in every contingency.

Stoppage Stoppage occurs when blockage occurs for the first time from some initial state.

Survivor   A survivor is the state that is selected at any stage as the result of backward induction.

Symmetric game   A symmetric game is a two-person normal (strategic)-form game, in which the payoff ranks along the main diagonal can be arranged so that they are the same for each player, and the payoff ranks along the off-diagonal are mirror images of each other.

Theory of moves (TOM)   The theory of moves describes optimal strategic calculations in normal-form games, in which the players can move and countermove from an initial state.

Threat outcome/strategy   A threat outcome is the Pareto-superior state that a threatener promises to implement, by choosing its threat strategy, if the threatened player also agrees to its choice. See also *Breakdown outcome/strategy*.

Threat power   In a two-person game that is repeated, threat power is the ability of a player to threaten a mutually disadvantageous outcome in the single play of a game to deter untoward actions in the future play of this or other games. See also *Compellent threat; Deterrent threat*.

Two-sidedness rule   The two-sidedness rule describes how players determine whether or not to move from a state on the basis of the other players' rational choices as well as their own. See also *Rationality rules; Rules of play*.

Undecidability   SB is undecidable if his strategy choices in a game, and the game's outcome, cannot be distinguished from the choices and outcome of a game in which SB is a human player, without supernatural abilities.

Undominated strategy   An undominated strategy is a strategy that is neither a dominant nor a dominated strategy.

Utility   Utility is the numerical value, indicating degree of preference, that a player has for an outcome.

Variable-sum game   A variable-sum game is a game in which the sum of the payoffs to the players in different states is not constant but variable, so the players may gain or lose simultaneously in different states.

Zero-sum game   See *Constant-sum (zero-sum) game*.

References

Ackerman, James S. 1974. "The Literary Context of the Moses Birth Story (Exodus 1–2)." In *Literary Interpretations of Biblical Narratives*, ed. Kenneth R. R. Gros Louis, James S. Ackerman, and Thayer S. Warshaw. Nashville, TN: Abingdon Press.

Allen, Woody. 2007. *Mere Anarchy*. London: Edbury; New York: Random House.

Amir, Rabah. 1995. "Endogenous Timing in Two-Player Games: A Counterexample." *Games and Economic Behavior* 9 (2): 234–237.

Anchor Bible: Genesis. 1964. Ed. E. A. Speiser. Garden City, NY: Doubleday.

Binmore, Ken. 2009. *Rational Decisions.* Princeton, NJ: Princeton University Press.

Bok, Sissela. 1978. *Lying: Moral Choice in Public and Private Life.* New York: Pantheon.

Brams, Steven J. 1975. "Newcomb's Problem and Prisoners' Dilemma." *Journal of Conflict Resolution* 19 (4): 596–612.

Brams, Steven J. 1976. *Paradoxes in Politics: An Introduction to the Nonobvious in Political Science.* New York: Free Press.

Brams, Steven J. 1977. "Deception in 2 × 2 Games." *Journal of Peace Science* 2 (Spring): 171–203.

Brams, Steven J. 1982. "Omniscience and Omnipotence: How They May Help—or Hurt—in a Game." *Inquiry* 25 (2): 217–231.

Brams, Steven J. 1994. *Theory of Moves.* Cambridge, UK: Cambridge University Press.

Brams, Steven J. [1980] 2003. *Biblical Games: Game Theory and the Hebrew Bible.* Rev. ed. Cambridge, MA: MIT Press.

Brams, Steven J. [1983] 2007. *Superior Beings: If They Exist, How Would We Know? Game-Theoretic Implications of Omniscience, Omnipotence, Immortality, and Incomprehensibility.* New York: Springer.

Brams, Steven J. 2011. *Game Theory and the Humanities: Bridging Two Worlds.* Cambridge, MA: MIT Press.

Brams, Steven J. 2018. "Resolving Solomon's Dilemma." Preprint.

Brams, Steven J., and Marek P. Hessel. 1983. "Staying Power in 2 × 2 Games." *Theory and Decision* 15 (3): 279–302.

Brams, Steven J., and Marek P. Hessel. 1984. "Threat Power in Sequential Games." *International Studies Quarterly* 28 (1): 15–36.

Brams, Steven J., and Christopher B. Jones. 1999. "Catch-22 and King-of the-Mountain Games: Cycling, Frustration, and Power." *Rationality and Society* 11 (2): 139–167.

Brams, Steven J., and D. Marc Kilgour. 2017. "Stabilizing Unstable Outcomes in Prediction Games." Preprint.

Brams, Steven J., and Frank C. Zagare. 1977. "Deception in Simple Voting Games." *Social Science Research* 6 (3): 257–272.

Brams, Steven J., and Frank C. Zagare. 1981. "Double Deception: Two against One in Three-Person Games." *Theory and Decision* 13 (3): 81–90.

Bruns, Bryan Randolph. 2015. "Names for Games: Locating 2 × 2 Games." *Games* 6 (4): 495–520.

Buber, Martin. 1958. *I and Thou.* 2nd ed. Trans. R. G. Smith. New York: Scribner's.

Bueno de Mesquita, Bruce 1996. "Counterfactuals and International Affairs: Some Insights from Game Theory." In *Counterfactual Thought Experiments in World Politics: Logical, Methodological, and Psychological Perspective*, ed. Philip E. Tetlock and Aaron Belkin, 211–229. Princeton, NJ: Princeton University Press.

Chimenti, Frank A. 1990. "Pascal's Wager: A Decision-Theoretic Approach." *Mathematics Magazine* 63 (5): 321–325.

Chwe, Michael Suk-Young. 2014. *Jane Austen, Game Theorist.* Princeton, NJ: Princeton University Press.

Cohen, Raymond. 1991. *Negotiating across Cultures.* Washington, DC: U.S. Institute of Peace.

Cowley, Robert, ed. 2000. *What If? The World's Foremost Military Historians Imagine What Might Have Been.* New York: G. P. Putman's Sons (Penguin).

Cowley, Robert, ed. 2001. *What If 2? Eminent Historians Imagine What Might Have Been.* New York: G. P. Putman's Sons (Penguin).

Cowley, Robert, ed. 2004. *What If of American History?* New York: Berkeley.

Dawkins, Richard. 2006. *The God Delusion.* Boston: Houghton Mifflin Harcourt.

Dennett, Daniel C. 2006. *Breaking the Spell: Religion as a Natural Phenomenon.* New York: Viking.

Dershowitz, Alan M. 2000. *The Genesis of Justice: Ten Stories of Biblical Injustice That Led to the Ten Commandments and Modern Law.* New York: Warner.

Diller, Jeanine, and Asa Kasher, eds. 2013. *Models of God and Alternative Ultimate Realities.* Dordrecht, Netherlands: Springer.

Ferejohn, John A. 1975. Personal communication (May 27).

Fishburn, Peter C. 1974. "Lexicographic Orders, Utilities and Decision Rules: A Survey." *Management Science* 20 (11): 1442–1471.

Fishburn, Peter C., and D. Marc Kilgour. 1990. "Binary 2 × 2 Games." *Theory and Decision* 29 (3): 165–182.

Fowler, James W. 1981. *Stages of Faith: The Psychology of Human Development and the Quest for Meaning.* San Francisco: Harper and Row.

Fraser, Niall M., and D. Marc Kilgour. 1986. "Non-Strict Ordinal 2 × 2 Games: A Comprehensive Computer-Assisted Analysis of the Possibilities." *Theory and Decision* 20 (2): 99–121.

Fraser, Niall M., and D. Marc Kilgour. 1988. "A Taxonomy of All Ordinal 2 × 2 Games." *Theory and Decision* 24 (2): 99–117.

Gardner, Martin. 1973. "Mathematical Games." *Scientific American* 229 (1) (July): 104–108.

Gardner, Martin (written by Robert Nozick). 1974. "Mathematical Games." *Scientific American* 230 (3) (March): 102–108.

Gilboa, Itzhak. 2009. *Theory of Decision under Uncertainty.* Cambridge, UK: Cambridge University Press.

Goldstein, Rebecca. 2006. *Incompleteness: The Proof and Paradox of Kurt Gödel.* New York: W. W. Norton.

Goodman, James E. 2013. *But Where Is the Lamb? Imagining the Story of Abraham and Isaac.* New York: Schocken.

Graves, Robert, and Raphael Patai. 1963. *Hebrew Myths: The Book of Genesis.* New York: McGraw-Hill.

Hamilton, Jonathan H., and Steven M. Slutsky. 1988. "Endogenizing the Order of Moves in Matrix Games." *Theory and Decision* 34 (1): 47–62.

Hamilton, Jonathan H., and Steven M. Slutsky. 1990. "Endogenous Timing in Duopoly Games: Stackelberg or Cournot Equilibria." *Games and Economic Behavior* 2 (1): 29–46.

Hanson, Norwood Russell. 1971. *What I Don't Believe and Other Essays.* Ed. Stephen Toulmin and Harry Woolf. Dordrecht, Netherlands: D. Reidel.

Hitchens, Christopher. 2007. *God Is Not Great: How Religion Poisons Everything.* New York: Hachette.

Hurly, S. L. 1994. "A New Take from Nozick on Newcomb's Problem and Prisoners' Dilemma." *Analysis* 54 (2): 65–72.

James, William. [1902] 1967. *The Writings of William James,* ed. John J. McDermott. Chicago: University of Chicago Press.

Jordan, Jeff. 2006. *Pascal's Wager: Pragmatic Arguments and Belief in God.* Oxford: Oxford University Press.

Kierkgaard, Søren. [1843] 1954. *Fear and Trembling.* Trans. Walter Lowrie. Princeton, NJ: Princeton University Press.

Kilgour, D. Marc, Mitali De, and Keith W. Hipel. 1987. "Conflict Analysis Using Staying Power." *Proceedings of the 1986 IEEE International Conference on Systems, Man, and Cybernetics,* Atlanta, GA.

Kilgour, D. Marc, and Frank C. Zagare. 1987. "Holding Power in Sequential Games." *International Interactions* 13 (2): 321–347.

Kolakowski, Leszek. 1972. *The Key to Heaven.* Trans. C. Wieniewska and S. Attanasio. New York: Grove Press.

Kolakowski, Leszek. 1982. *Religion*. New York: Oxford University Press.

Küng, Hans. 1980. *Does God Exist? An Answer for Today*. Trans. E. Quinn. New York: Doubleday.

Landsberg, P. T. 1971. "Gambling on God." *Mind* 80 (317): 100–104.

Lewis, David. 1985. "Prisoners' Dilemma Is a Newcomb Problem." In *Paradoxes of Rationality and Cooperation*, ed. Richmond Campbell and Lanning Sowden, 251–255. Vancouver: University of British Columbia Press.

Mailath, George J., Larry Samuelson, and Jeroen Swinkels. 1993. "Extensive Form Reasoning in Normal Form Games." *Econometrica* 61 (2): 273–302.

Mailath, George J., Larry Samuelson, and Jeroen Swinkels. 1994. "Normal Form Structures in Extensive Form Games." *Journal of Economic Theory* 64 (2): 325–371.

McShane, Paddy Jane. 2014. "Game Theory and Belief in God." *International Journal for Philosophy of Religion* 75 (1): 3–12.

Miles, Jack. 1995. *God: A Biography*. New York: Vintage.

Muzzio, Douglas. 1982. *Watergate Games: Strategies, Choices, Outcomes*. New York: New York University Press.

Nozick, Robert. 1969. "Newcomb's Problem and Two Principles of Choice." In *Essays in Honor of Carl G. Hempel*, ed. Nicholas Rescher, 114–146. Dordrecht, Netherlands: D. Reidel.

Nozick, Robert. 1997. "Reflections on Newcomb's Problem." In *Socratic Puzzles*, 74–84. Cambridge, MA: Harvard University Press.

O'Neill, Barry. 1999. *Honor, Symbols, and War*. Ann Arbor: University of Michigan Press.

Pascal, Blaise. [1670] 1950. *Pensée*. Trans. H. F. Stewart. New York: Pantheon.

Pawlowitsch, Christina. 2012. "Meaning, Free Will, and the Certification; of Types in a Biblical Game." Preprint.

Prophets, The. 1978. Philadelphia: Jewish Publication Society.

Rapoport, Anatol, and Melvin J. Guyer. 1966. "A Taxonomy of 2 × 2 Games." *General Systems: Yearbook of the Society for General Systems Research* 11:203–214.

Rapoport, Anatol, Melvin J. Guyer, and David G. Gordon. 1976. *The 2 × 2 Game*. Ann Arbor: University of Michigan Press.

Rescher, Nicholas. 1985. *Pascal's Wager: A Study of Practical Reasoning in Philosophical Theology*. South Bend, IN: University of Notre Dame Press.

Rescher, Nicholas. 2001. *Paradoxes: Their Roots, Range, and Resolution*. Chicago: Open Court.

Robinson, David, and David Goforth. 2005. *The Topology of 2 × 2 Games: A New Periodic Table*. New York: Routledge.

Rosenthal, Robert W. 1991. "A Note on Robustness of Equilibria with Respect to Commitment Opportunities." *Games and Economic Behavior* 3 (3): 237–242.

Sarna, Nahum M. 1970. *Understanding Genesis: The Heritage of Biblical Israel*. New York: Schocken.

Stackelberg, Heinrich von. 1952. *The Theory of the Market Economy*. Trans. Alan Peacock. London: William Hodge. Originally published in German: *Marktform und Gleichgewicht*. Berlin: J. Springer, 1934.

Styron, William. 1979. *Sophie's Choice*. New York: Random House.

Swinburne, Richard. 1981. *Faith and Reason*. Oxford: Clarendon.

Tarar, Ahmer. 2018. "A Game-Theoretic Analysis of Pascal's Wager." *Economics and Philosophy* 34 (1): 31–44.

Taylor, Alan D., and Allison D. Pacelli. 2008. *Mathematics and Politics: Strategy, Voting, Power, and Proof*. 2nd ed. New York: Springer.

Tetlock, Philip E., and Aaron Belkin, eds. 1996. *Counterfactual Thought Experiments in World Politics: Logical, Methodological, and Psychological Perspectives*. Princeton, NJ: Princeton University Press.

Torah: The Five Books of Moses, The. 2nd ed. 1967. Philadelphia: Jewish Publication Society.

van Damme, Eric, and Sjaak Hurkens. 1996. "Commitment Robust Equilibria and Endogenous Timing." *Games and Economic Behavior* 5 (2): 290–311.

von Neumann, John, and Oskar Morgenstern. [1944] 1953. *Theory of Games and Economic Behavior*. 3rd ed. Princeton, NJ: Princeton University Press.

Weingast, Barry R. 1996. "Off-the-Path Behavior: A Game-Theoretic Approach to Counterfactuals and Its Implications for Political and Historical Analysis." In *Counterfactual Thought Experiments in World Politics: Logical, Methodological, and Psychological Perspectives*, ed. Philip E. Tetlock and Aaron Belkin, 230–243. Princeton, NJ: Princeton University Press.

Wiesel, Elie. 1977. *Messengers of God: Biblical Portraits and Legends*. New York: Pocket Books.

Willson, Stephen J. 1998. "Long-Term Behavior in the Theory of Moves." *Theory and Decision* 45 (3): 201–240.

Writings, The. 1982. Philadelphia: Jewish Publication Society.

Zagare, Frank C. 1984. "Limited-Move Equilibria in 2 × 2 Games." *Theory and Decision* 16 (1): 1–19.

Zagare, Frank C. 2011. *The Games of July: Explaining the Great War*. Ann Arbor: University of Michigan Press.

Zeager, Lester A., Richard E. Ericson, and John H. P. Williams. 2013. *Refugee Negotiations from a Game-Theoretic Perspective*. Dordrecht, Netherlands: Republic of Letters.

Index

Aaron, 30, 131, 133–135, 144–146, 149–150, 153
Abel, 98, 109–117, 125, 132, 173–174
Abigail, 176
Abraham, 59, 73, 89–102, 104–109, 127, 129–130, 132–133, 145, 152, 171–173, 176
Ackerman, James S., 129
Adam and Eve, 75–89, 98, 110, 114, 128, 164
Adultery, 53, 176
Aesop's principle, 15n3
Afterlife, 13
Agag, 120
Age of Enlightenment, 32
Age of Reason, 32
Agnosticism, xi–xii, 11, 16, 18–20, 29–30, 168, 177
Allen, Woody, 96n7
Amalek, 120
Amalekites, 120, 122, 143
Amir, Rabah, 41n2
Ammonites, 89, 99–102, 120, 172
Amsel, Larry, xiv
Anchor Bible: Genesis, 95n6
Asimov, Isaac, 66, 73
Atheism, xi, 18, 29

Backward induction. *See* Theory of moves (TOM), backward induction in
Bathsheba, 176
Belief Games, 21–54, 62, 70, 76, 80, 81n4
Belkin, Aaron, 98n11
Betrayal, 133
Bible, Hebrew, books of, xiiin4, 91, 103, 148, 151, 156–157
Binmore, Ken, 15n
Biology, evolutionary, 4

Blame Game, 109, 114–118, 125, 174, 180
Bluff, 97
Bok, Sissela, 164n
Bradley, James, xiv
Brams, Steven J., xii, 1n, 4, 9n, 18n4, 22n2, 26n9, 28, 29n15, 41n2, 42n4, 43, 53, 55n1, 63n5, 67n8, 69n9, 70–71, 76n1, 85, 91n2, 92n4, 97n8, 102, 129, 155n1, 163n3, 164, 179, 180
Bruns, Bryan Randolph, 179n1
Buber, Martin, 9
Bueno de Mesquita, Bruce, 98n11
Burning bush, 130

Cain, 98, 109–118, 125, 128, 132, 148, 157, 173–174, 176
Caleb, 149–150
Canaan, 102, 118, 149–151, 154
Chicken, game of, 56, 66n7, 70–73, 113n4, 180, 185
Chimenti, Frank A., 10
Chosen people. *See* Israelites
Chwe, Michael, 4
Cohen, Raymond, 21n1
Collective consciousness, 32
Common knowledge, 23, 185. *See also* Information, complete/incomplete
Concern Decision, 10, 16–20, 36, 168
Conditional cooperation, 63n5
Counterfactual, 98n11, 165
Constant-sum game. *See* Game, of total conflict
Constraint Game, 75–81, 84–85, 87–88, 110, 128, 180
Covenant, 148
Creation, 6–8, 75, 98, 118
Crusades, 32

Index of Biblical Passages

Printed in the United States
by Baker & Taylor Publisher Services